Transportation Systems: A Practical Approach

Transportation Systems: A Practical Approach

Edited by
Samuel Morgan

STATES
ACADEMIC PRESS
www.statesacademicpress.com

Published by States Academic Press,
109 South 5th Street,
Brooklyn, NY 11249, USA

ISBN: 978-1-63989-530-4

Cataloging-in-Publication Data

Transportation systems : a practical approach / edited by Samuel Morgan.
p. cm.
Includes bibliographical references and index.
ISBN 978-1-63989-530-4
1. Transportation engineering. 2. Transportation. 3. Intelligent transportation systems. 4. Traffic engineering.
I. Morgan, Samuel.
TA1145 .T73 2022
629.04--dc23

For information on all States Academic Press publications
visit our website at www.statesacademicpress.com

STATES
ACADEMIC PRESS

Contents

Preface

It is often said that books are a boon to mankind. They document every progress and pass on the knowledge from one generation to the other. They play a crucial role in our lives. Thus I was both excited and nervous while editing this book. I was pleased by the thought of being able to make a mark but I was also nervous to do it right because the future of students depends upon it. Hence, I took a few months to research further into the discipline, revise my knowledge and also explore some more aspects. Post this process, I begun with the editing of this book.

Transportation systems can be defined as systems for moving persons or goods consisting of the vehicle, the guideway and the operation plan. They are a fundamental part of logistics and planning whenever vehicles are used to move people or items from one place to another. Transportation systems include all forms of transport, from cars and buses to boats, aircraft and even space travel. Public transport systems are available to the general public, may require a fare, and run at scheduled times. Transportation systems make use of route management and effective timetabling to reduce transport costs and improve delivery time. This book discusses the fundamentals as well as modern approaches of transportation systems. From theories to research to practical applications, case studies related to all contemporary topics of relevance to this field have been included in it. This book, with its detailed analyses and data, will prove immensely beneficial to professionals and students involved in this area at various levels.

I thank my publisher with all my heart for considering me worthy of this unparalleled opportunity and for showing unwavering faith in my skills. I would also like to thank the editorial team who worked closely with me at every step and contributed immensely towards the successful completion of this book. Last but not the least, I wish to thank my friends and colleagues for their support.

Editor

1

The Possible Role of Large-Scale Sewage Plants in Local Transport

Attila Bai and Zoltán Gabnai

Abstract

Large-scale sewage plants in large cities are suitable for the production of large quantities of biogas, using economically viable biogas upgrading technologies and generally available public transport fl ets of a sufficie t number of local buses, as well as municipal vehicles. The conditions for the sale of locally produced CNGs do not depend on gas suppliers, they can be very well integrated with local waste management, and the local emission reductions occur in the inner city, where air pollution is the most serious problem. At the same time, the cogeneration solution currently of decisive importance for wastewater plants is more economically and environmentally advantageous in the production of biomethane. The consumption of heat and electricity by these plants is signifi ant and must be supplied through the purchase of biomethane. However, for the local authority, when converting diesel buses, compressed biomethane (CBM) offers much greater savings, so at the municipal level, the process is economically profitable. The short-term spread of CBM (due to the small number of filling stations) is bound to local systems. If more and more cities operated a similar system (allowing refuelling within a few dozen kilometres), it would be expected that passenger cars would also be more widespread.

Keywords: sludge management, biomethane, local transport, sustainability, economics

Introduction

The EU-28 2016 energy use was 1147 million tonnes of oil equivalent (MTOE), 17% of which was from renewable energy sources; the proportion of renewable fuels was 7.1% [1]. The proportion of renewables was 9.5% in 2006, so renewable energies have cca. 8% growth per year in average during the last decade in the process of energy production [2].

However, the average levelized cost of electricity (LCOE) of fuel generation and biogas electrical power generation is still much higher than the reference [3], and therefore without fi ancial support, it is not competitive with oil and other highly renewable electricity generation technologies (solar panels, geothermal energy, hydropower, solid-fi ed biomass power plants).

Biogas production is also a remarkable process in terms of the environment and energy production, but its effici cy, with new raw materials, technologies and markets, can be signifi antly increased. Due to the capital requirements and the uncertainty of innovative solutions, the establishment of biogas plants and their operation and the research conducted in this area are signifi antly supported by all EU member states. The dominant product of current biogas plants is green electricity, which is supported by competent power suppliers at a subsidised price and is acquired as a compulsory purchase by the relevant electricity suppliers, details of which are given for the EU-27, the EFTA countries and countries waiting to join in [4]. The feeding of biomethane purifi d with biogas to the natural gas pipeline is regulated by stringent standards in the EU member states, according to Szunyog [5]. However, gas with a lower methane content than natural gas may also be utilised as a propellant, although its compression and transport costs are larger, and the range is smaller than with compressed natural gas (CNG) [6].

Conversion of biogas will always increase investment and operational costs, and the energy effici cy of the process will be reduced, but it will produce more valuable, versatile and marketable main products (electricity, biomethane).

According to Hakawatia et al. [7], on the basis of 49 different biogas transformation technology studies, the overall effici cy of the process ranges from 16 to 83% in the case of direct burning of biogas and 8 to 54% for cogeneration (for electricity and waste heat coupled production), while when producing biomethane for fuel, it varied from 4 to 18%. If the electricity generated by cogeneration is used in electric vehicles, the effici cy of propellant use can be increased to 33%.

Most of the wastewater plants produce electricity and waste heat from biogas. The reason for this is clear from direct heat utilisation: it is almost impossible to use biogas exclusively for heating purposes in larger-sized plants and during summer time. On a large scale, however, the summer utilisation of waste heat generated during the cogeneration process is also problematic. In order to avoid the problems of heat utilisation, in the case of large plants, biomethane can be considered as an alternative to biogas purifi ation and the utilisation of by-product carbon dioxide in the algae sewage system. In our chapter, we would like to point out that the inclusion of biogas from wastewater treatment plants in large cities in vehicles involved in local transport (buses, taxis, public utility vehicles) can also be an environmentally and economically promising alternative, of which we can already fi d many well-functioning examples.

The transport sector, CNG and the significance of compressed biomethane (CBM)

Traffi is one of the most signifi ant sectors in the EU-28, with around €651 billion in gross value added a year at basic prices (5% of total), with 11.2 million employees (5.2% of total), 6602 billion passenger km (on average around 12,962 km per person, of this 8.2% with buses and coaches) and about 1183 million tonnes of CO2 equivalent (30.7% of total). Private households in the EU-28 spent EUR 230 billion on transport services (e.g. bus, train, plane tickets). It should be highlighted that changing consumer attitudes from fuel to more environment-friendly way of transport may help to promote the spreading of sludge-based transport fuels, too [8].

Globally, in 2014 transport was responsible for 23% of total CO2 emissions from fuel combustion, and road transport was responsible for 20% [9].

At the same time, bus transport is one of the safest modes of transport: only 126 out of the 26,134 traffi deaths in 2015 occurred on buses.

Theoretically the existence of good public transport can deter car ownership. The paper by Cullinane and Cullinane [10] asserts, however, that once a car has been acquired, there is a tendency for it to be used irrespective of how good the public transport is.

Natural gas engine technology is already well established, and millions of vehicles using natural gas are in operation worldwide and suitable for using CBM.

While at the turn of the millennium, a million registered vehicles around the world were powered by CNG;

by 2015 this figu e had already increased to 22.3 million, an average annual increase of 22%, well above the growth in the total fl et of cars. Typically, most of the gas-fuelled vehicles (Iran, 4 million; Pakistan, 3.7 million; Argentina, 2.5 million; India, 1.8 million) [11] are found in low-income countries with a high population density. In the EU, 3345 refuelling stations provide for the operation of about 1.3 million gas-powered vehicles with an annual gas consumption of around 5 billion cubic metres. For the operation and further spread of this existing transport infrastructure, CBM could make a signifi ant contribution [12].

The spread of biogas plants in the EU has been very rapid over the past decade: rising from 6227 to 17,662 between 2009 and 2013. Growth in farms was primarily signifi ant (12,496 plants), the number of sewage plants was 2838 and the remainder were landfill waste plants. The number of plants has stagnated since 2015, but the installed electric capacity has increased further; currently it is 9985 MW [13]. Anaerobic digestion is a key technology for the treatment of large volumes of biowaste [14].

The energy signifi ance of biogas is underlined by the fact that in the EU, the amount of biogas produced in 2015 reached 18.4 billion normal cubic metre (Nm^3), replacing 4% of natural gas consumption [15]. The amount of biomethane fed to the natural gas pipeline reached 1.5 Mrd m^3 (mainly the Netherlands), while the amount of biomethane used as propellant is considerably less, at 160 M m^3 (Sweden, 113 million m^3; Germany, 35 million m^3; Norway, 10 million m^3; Iceland, 2 million m^3; Finland, 0.2 million m^3) [16].

However, the use of purifi d biogas for transport in some countries is rapidly expanding: over 30 cities in Sweden power their municipal buses with biogas, which is also used by taxis and sanitation companies [17].

There are 247 biomethane plants around the world and around 80 in the EU. Their technology and the biomethane produced are characterised by the data in **Table 1**.

Regarding environmental performance, methane loss is of great importance, as methane is a greenhouse gas 21 times stronger than CO_2 [19]. As clarifi d by Beil and Beyrich [20], pressurised water scrubbing (PWS) is one of the best solutions in terms of effici cy and environmental performance.

Cleaning technology	Number of plants	Biomethane CH4 content (%)	Specific operating costs (USD/Nm3 biomethane)
Water wash	107	96.1	0.18
Pressure change adsorption	55	95.8	0.34
Chemical (amino) absorption	53	94.6	0.23/0.38*
Membrane	22	90.3	0.16/0.30*
Genosorb©	12	96	n.a.
Cryogen	1	88	0.59

**Hydrolysis purification.*
Source: Own construction based on Yang et al. [18].

Table 1.
Biogas cleaning technology and the cost of cleaning.

We need to grow food on even less land, with less water, using less energy, fertilizer and pesticide than we use today for feeding more and more people [21], so the energy use of wastes (e.g. sewage) instead of plants can be taken as one of the most important reserves of land management; with the use of them, the net use of land declines [22].

The environmental and technical characteristics of compressed natural gas (CNG)/compressed biogas (CBG) applications

Among the hydrocarbons known as energy carriers, methane is the simplest formulation molecule, with the highest hydrogen-to-carbon ratio. Th s feature allows it to achieve uniquely clean combustion among internal combustion engines (**Table 2**).

If the source material of the CNG is biogas, it can save a further 13 kg CO_2e/t when supplied with organic fertilisers, by replacing artific al fertilisers [23].

Th CO_2 emissions of biogas plants throughout their life cycle also depend to a large extent on the composition of organic matter. Fuchsz and Kohlhéb [27] calculate that cogeneration-based animal fertiliser-based colonies in electricity production and in all environmental categories (GHG, eutrophication potential,

acidifi ation potential) are less effective than biogas plants which (also) use energy crops. Th s is equally true for both the period of setting up the plant and the operational period. The CO_2 emissions of all three examined types were below the average emissions of electricity currently produced in a natural gas-fuelled power plant. At the same time, Bordelanne et al. [26] explicitly analysed the life cycle of sewage waste, fi ding that with the fermentation of municipal wastes, energy plants also produce 7–10% less greenhouse gas emissions; thus, the biomethane production of fuel in our test is the most environmentally friendly compared to other biogas raw materials.

Examining the various pollutants, CNG CO and NOx emissions were 50–80% of those for EURO 4 petrol vehicles [26], while CH values were close to the same values.

The technology of diesel buses also signifi antly influences the environmental impact of CNG/CBM. According to Ryan and Caulfi ld's [28] estimates, the use of CNG in the case of all diesel particulates used in EURO 2–4 buses was reduced to a minimum of 70% for all pollutants and 100% for SO2 and heavy metal emissions. The GHG emissions of the CNG tested were also 7% better than the CBM. Compared to EURO 5 buses, the emissions of CNG buses were at least 50% more favourable for most air pollutants, but the CO and non-methane volatile organic compound (NMVOC) values for diesel were slightly more favourable.

GHG savings over the entire life cycle (%)*				
Biomethane raw material	Maize silage	Fertiliser	Municipal solid waste	
Compared to diesel	66	96	95	
Compared to petrol	70	97	96	
GHG emissions produced during engine use (kg/GJ)				
CNG/CBG	Petrol	Diesel	Biodiesel	Bioethanol
56**/65***/74****	74**/93****	73**/93****	76***	71***

*Sources: Own construction based on * [23] ** [24], *** [25], **** [26] (figures refer to life cycles).*

Table 2.
GHG savings with CNG.

Regarding the emission factors of petroland diesel-fuelled cars and buses, it can be stated with great certainty that bus transport is the best option, in almost any of harmful gases with respect to g/pkm (passenger km), assuming that the car is not full of passengers. The emission factors strongly depend on size and utility of the given vehicles. CO and CH4 emission is higher in petrol-fuelled cars, while NOx emission is stronger in diesel-fuelled vehicles including buses (**Figure 1**). It should be noted that CH_4 and CO emission can reach outstanding values regarding motorcycles.

Methane content also affects the emission of pollutants. Lim et al. [31] examined gases composed of 82–98% methane and found that total hydrocarbons (THC), CO, NOx and CO_2 emissions increased, while volatile organic compound (VOC) emissions decreased with an increased CH4 content in the fuel.

Because the activation energies of high carbon-numbered hydrocarbons (HCs) (e.g. ethane or propane) are lower than those of low carbon-numbered HCs (methane), the combustion effici cies of ethane and propane are greater than that of methane. Thus, a fuel with higher methane content has more incomplete combustion and/or poor oxidation of unburned hydrocarbons, resulting in higher emissions of HC [31]. The level of this, according to measurements taken by Subramanian et al. [32], was 12.5%.

Considering biogas and natural gas, NOx emissions were lower for biogas than for the natural gases.

Biomethane used as a biofuel produces emission savings which are 73–82% of the base values (GVA) used for the EU's sustainability legislation. Compared to biofuels

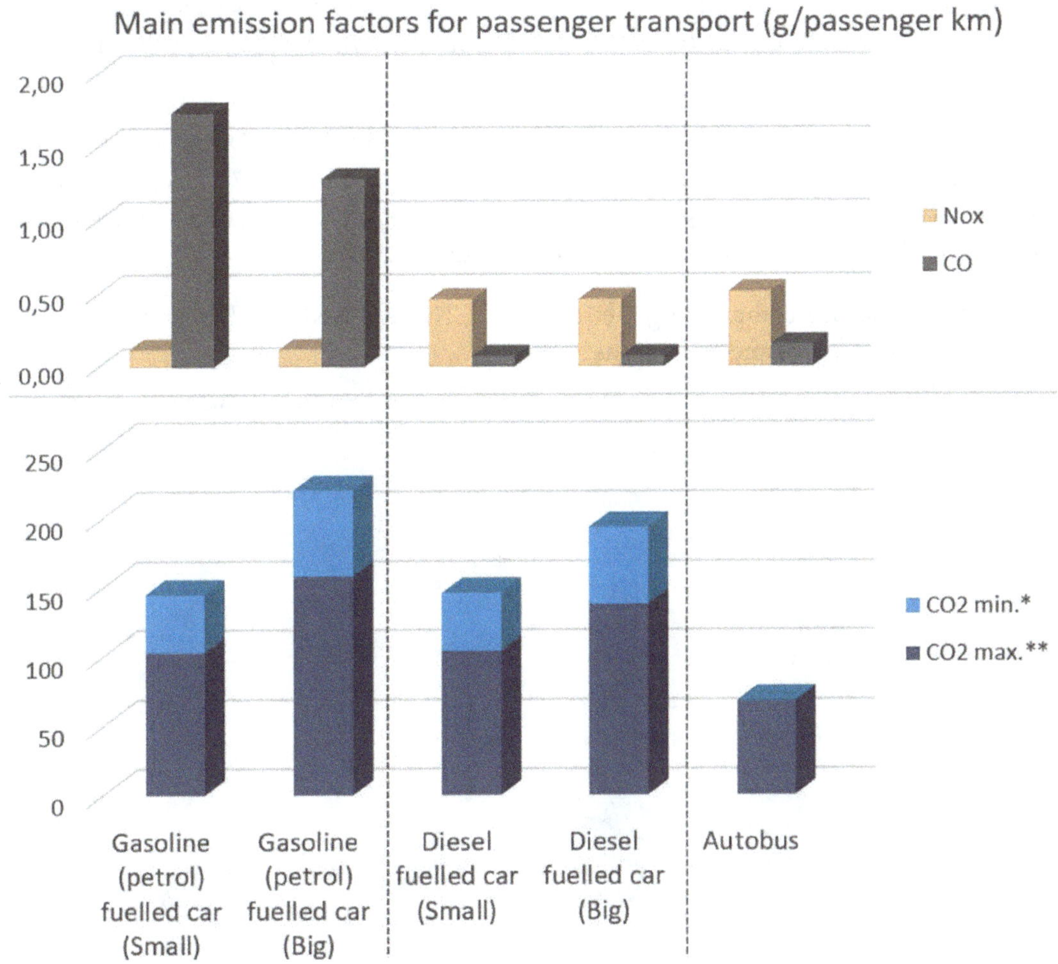

Figure 1.

Main emission factors for passenger transport (g/pkm). Source: Own construction based on [29, 30].

produced from other raw materials, this is very favourable because beet ethanol produces 52%, wheat ethanol 16–69%, maize ethanol 49% and rapeseed biodiesel 38% (Directive 2009/30 EC) [33].

According to data from Goulding and Power [34], the biomethane energy yield from grass silage is 2.5–3.3, which corresponds to 67–78 GJ of annual energy surplus per hectare. The figu es for wastewater with the same biogas yield are between 3.5–4 and 85–90 GJ/year. These are also well above those of most biofuels.

The specific pollutant emissions per capita of public transport are much lower than those of car transport, and this is especially true if the fuel itself is environmentally friendly. Since many trips are local, the analysis by the Department for Transport (UK) shows that 44% of all CO_2 emissions from cars come from journeys of between 5 and 25 miles [35]. Although the pollutant emissions of buses per passenger kilometre are higher than those for trains and trams (**Figure 2**), their energy consumption is nearly identical and much better than for individual transport [29].

The use of raw biogas in CNG vehicles has been investigated. These tests have shown that raw biogas (not upgraded) can be used as a fuel, if blended with natural gas. In fact, the use of raw biogas can be envisaged in dedicated CNG engines, if new engine technologies (lean CNG combustion) are developed. In such a case, natural gas can be blended with up to 70% volume of non-upgraded biogas.

Tests by Bordelanne et al. have shown that raw (not upgraded) biogas can be used as a fuel, only if blended with natural gas, mainly in CNG engine types (lean CNG combustion). The biogas proportion of natural gas can be a maximum of 70% [26].

When comparing two identical brands and types of waste collection vehicles, Domanovszki [36] concluded that the average noise level measured by microphones located at a distance of 7 m from vehicles is 71 dB in the diesel engine and 66 dB in the gas engine version, which is about three times the noise load. Gas fuels do not contain anti-knock additives due to their high-octane number.

There are two technologies for compressing methane gas: CNG and LNG. In Hungarian and EU practice, among gas-fi ed propellants, propane-butane gas

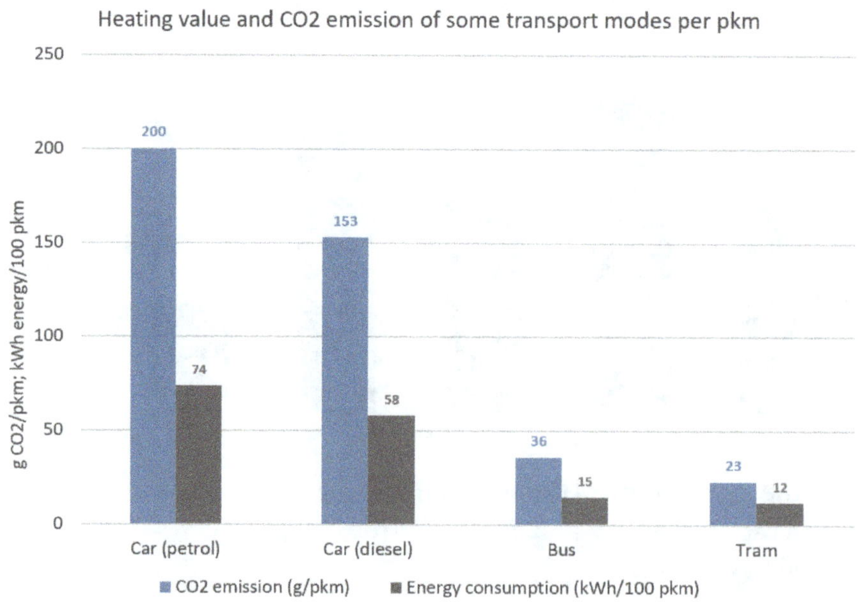

Figure 2.
Heating value and CO_2 emission of some transport modes per pkm. Source: Own construction based on [30].

(LPG) is the most widespread, while in the United States, it is compressed natural gas (CNG), and the infrastructure for the latter is perfectly suited to biomethane utilisation. CNG is, in theory, subjected to 200 bar pressure when put into steel or composite tanks. With LNG technology, a higher energy density (55 MJ/kg, 24–26 MJ/Nm3) can be obtained by liquefaction at -161.6°C at atmospheric pressure. It can then be filled into well-insulated containers and stored under low pressure. Because of the higher calorific value, LNG is more suitable for long-distance traffic than CNG [36]. If biogas is the raw material, the CBM operation will have a combined efficiency of 15–18%, and LMS (liquified biomethane) an efficiency of 14–17%, but both significantly outperform the overall efficiency of the liquid fuel systems (4–13%) [7].

Economic characteristics of production and use

Large-scale sewage plants in large cities are suitable for the production of large quantities of biogas, using economically viable biogas upgrading technologies and generally available public transport fleets of a sufficient number of local buses, as well as municipal vehicles. The conditions for the sale of locally produced CNGs do not depend on gas suppliers, they can be very well integrated with local waste management and the local emission reductions occur in the inner city where air pollution is the most serious problem.

At the same time, it is not clear from the point of view of sewage plants whether it is the production of biomethane or the cogeneration solution—which is currently more important—which is more viable economically and in terms of harmful emissions. Here, it must be taken into account that the consumption of heat and electricity by these plants is significant, and this must be purchased when producing biomethane.

The economic and environmental approach should take into account not only the substituted energy source but also the cost and emissions of fossil fuels bought because there is no cogeneration (including the electricity and heat needed for self-consumption at sewage plants).

In this section we present the investment and operating costs of CBM production and the specific aspects of economic evaluation.

Investment and operating costs

The expected level of the biomethane plant's investment costs is greatly influenced by technology and size. Goulding and Power [34] provide the following equation for the average of the investment costs of biomethane plants operating with agricultural raw materials, with a 95.6% probability:

$$C ¼ 1066:2x^{0:8455}$$

where C is the investment cost (€/t yearly raw material) and x is the processed raw material quantity (t/year).

The equation—with a 91% degree of confide ce ($C = 21080x^{0.5367}$)—also shows that, in the case of larger dimensions, from an economies of scale perspective, it is advisable to use biomethane instead of cogeneration technologies. At the same time, at bigger dimensions cogeneration technology also has economies of scale. According to Patrizio et al. [37], the proportion of available heat and electricity between an operating capacity of 300 and 200 kWe increases from 58 to 70%, considering a value of 25 and 11% for plant self-consumption of electricity and for heat to maintain the mesophilic process.

Yang et al. [18] also explored the cost of investment in various cleaning technologies at various farm sizes, which support the economies of scale of large plants:

- 1000 Nm^3/h in crude (input) biogas capacity: 1.6–2 million USD
- 250 Nm^3/h in crude (input) biogas capacity: 0.7–1.1 million USD

The distribution of investment costs by the same source can be characterised by the following average numbers:

- Cleaning: 40–45%
- Compression: approx. 5%
- Distribution: 50–55%

Most of the investment costs are related to distribution, the proportion of which is largely dependent on the method used to transport methane [25]:

- CNG lorry transport: 12 €/GJ biomethane
- LNG lorry transport: 7 €/GJ biomethane
- CNG pipeline delivery: 5 €/GJ biomethane

The cost price of methane produced is also largely dependent on the size of the wastewater treatment plant, primarily because of the signifi ant proportion of fi ed costs (**Table 3**).

Regarding the economic indicators of the wastewater plants, it can be said that the existing rotting equipment and the larger size of the plant typically offer more favourable costs, thus allowing faster returns.

It should be noted that a reduction in the cost of biomethane can also be achieved by using carbon dioxide resulting from purifi ation (in some cases in pure form), which has a signifi ant increase in yields in greenhouses or algae plants. Algae can also be easily integrated into sewage treatment or can be used in bioenergy production and are capable of doubling their yield (up to 400 t/ha/year) with inorganic nutrients in wastewater and carbon dioxide [39, 40]. With the use of digested effluent *Paulownia tomentosa* [41] or *Sida hermaphrodita* [42], plantations may support the aim to meet the growing needs for site remediation and biomass production.

Source	Farm-sized plant (250 Nm^3/h)	Medium-sized plant (500 Nm^3/h)	Large operation (1000 Nm^3/h)
[25]	1.1	0.8	0.7
[38]	0.8–1*	0.7–0.9*	0.6–0.7*
**Assuming 0–100% of purchased raw material.*			

Table 3.
Cost of biomethane based on farm size and raw material (EUR/kg).

Patrizio et al. [37] examining the influence of the low market value of CO_2 found that starting from a carbon price of 15 EUR/tCO_2, the cogeneration option is preferable if plants are located in the proximity of existing district heating infrastructure. CNG plants are only competitive starting at a carbon price of 70 EUR/tCO_2 in areas with high feedstock availability, when the fi st upgrading plant for CNG production is introduced into the optimum mix.

Finally, it should be noted that the consumption of diesel oils tested by Farkas et al. [43] in Hungary showed a 5% difference in the same vehicles. Th s is important because the substitution value of the biomethane and its environmental savings depend greatly on the quality of the substituted oil product.

The European market

Market prices are examined for the most important biomethane-producing countries and for Hungary. Including this latter country is justifi d by the fact that the case study we presented is also Hungarian. CNG is cheaper than its competitors —petrol and diesel—in all countries, not only in terms of unit price but also petrol and gas oil equivalents, with a typical price difference of between 33–57% and 25–48%. Interestingly, there is a large difference between the two most widely regarded model countries in the EU—Italy and Sweden; in the former it is consumed for reasons of economy, and in the latter because of the environmental consciousness there. CNG prices in the countries surveyed show that differences can be more than double, but petrol and gas oil prices are much more balanced (**Table 4**).

Rebuilding of passenger cars to use alternative fuels involves additional costs and changes in consumption, the values of which are:

- CNG conversion, approx. 1.600 EUR; extra consumption 0–10% [11]
- LPG conversion, approx. 1.000 EUR; extra consumption 5–20% [11]

Country	Fuel consumer price					CNG price as percentage (%) of other propellants	
	CNG EUR/kg	EUR/l petrol equivalent	Petrol EUR/l	EUR/l diesel equivalents	Diesel EUR/l	Petrol	Diesel
Bulgaria	0.71	0.52	1.2	0.59	1.23	43	48
Finland	1.34	0.98	1.54	1.11	1.47	64	75
Germany	1.07	0.79	1.53	0.88	1.4	51	63
Hungary	1.12	0.82	1.22	0.92	1.34	67	69
Italy	0.99	0.73	1.66	0.82	1.56	44	52
Sweden	1.87	1.37	1.48	1.54	1.58	93	98
Average	1.18	0.87	1.44	0.98	1.43	60	68
Relating to buses			1.08		1.07		

Units used: CNG, 43.6 MJ/kg; petrol, 32 MJ/l; diesel, 36 MJ/l [46].
Source: [44–46].

Table 4.
CNG, gasoline and diesel prices for key EU countries and two Central and Eastern European countries (October 2018).

- E85 conversion, EUR 300; extra consumption 20–25% [11]
- Vegetable oil conversion, from 1500 to 5000 EUR; excess consumption 10% [47]
- Biodiesel conversion, from 1000 to 4000 EUR; surplus consumption 10% [48]

For buses, conversion to CNG operation costs € 30,000–€ 50,000. CNG-powered cars are about € 3000 to 5000 more expensive than their petrol and gas oil counterparts of the same brand and with the same parameters.

It is clear that all alternatives to CNG involve increased consumption, ideally off et by the more favourable price of the alternative fuel.

Comparison of energy self-supply and biomethane production of sewage plants

In our case study, in order to quantify energy possibilities, we conducted our calculations for a nutrient-containing wastewater treatment plant using a anaerobic fermentation technology for a population equivalent of 100,000 inhabitants (equivalent to an adult inhabitant's wastewater), which was considered as average in Hungary. Based on our conception and data collection - which were partly promoted by the Higher Education Institutional Excellence Programme of the Ministry of Human Capacities in Hungary (No. 20428-3/2018/FEKUTSTRAT) - for the use of biogas produced during fermentation, we present two possibilities in this section:

- Cogeneration energy production and then own consumption (and, if possible, the sale of surplus heat energy)
- Biomethane production and sales: supply to the natural gas network or use as fuel

Basic data of the 100,000 household equivalent wastewater plants:

- Amount of treated wastewater: 13,000 m^3/day
- Biological oxygen demand (BOD) content: 6000 kg BOD/day (BOD, biological oxygen demand; oxygen demand for aerobic removal of organic matter from the aqueous phase through microorganisms)
- Specific iogas yield (per m^3 treated wastewater): 0.93 kWh/m^3
- CH_4 content of biogas: 65% (heating value: 6.5 kWh/m^3)
- Gas engine loss: 15%
- Heat energy/electricity ratio for power generation: 60%/40%
- Non-residential electricity average price in EU-28: 0.112 EUR/kWh
- Non-residential natural gas price in EU-28: 0.0078 EUR/MJ

Source: [49–51].

The amount of biogas generated by anaerobic technology, using the above (average) technological characteristics:

- 13,000 m^3 * 0.37 kWh/m^3 = 1860 Nm^3/day.

Allowing for a 15% loss of biogas above, a biogas calorific value of 6.5 kWh/m3 and the energy consumed in energy production, and calculating using 60% heat and 40% electricity ratios, the quantity of the fi al product with the two technologies examined is shown below.

Expected revenue from cogeneration

Electricity generation

- Quantity of electrical energy that can be produced: 0.37 kWh/m^3 * 13,000 m^3 = 4837 kWh/day
- Consumption of electricity at the plant (Kárpáti [52] and authors' own calculations, allowing for 0.59 kWh/m^3 sewage sludge): 13,000 m^3 sewage * 0.59 kWh/m^3 = 7709 kWh/day
- Electricity balance: -2872 kWh/day (electricity self-suffici cy: 63%)
- Savings: 4837 kWh/day * 0.112 EUR/kWh = 542 EUR/day = 197,745 EUR/year

Thermal power generation

- Quantity and value of heat energy to be produced: 0,558 kWh/m^3 * 13,000 m^3 = 7254 kWh/day = 26,114 MJ/day
- The heat energy consumption of the plant (assuming a heat demand of 0.40 kWh/m^3): 5140 kWh/day = 18,500 MJ/day
- Heat energy surplus: 7614 MJ/day
- The amount of heat energy self-supply: in principle 91%

However, in practice, heat utilisation beyond the plant's own heat demand is problematic, especially in the summer, and in addition, the heat energy consumption is also lower. For district heating purposes, depending on the length and insulation of the piping system, 10–15% heat loss can be expected. If the remainder of the winter heat surplus is fully utilised by the district heating system and the summer hot water demand is considered, then about 55–70% of the heat generated can be utilised. In the following we calculate on an assumption of 60%:

Average value of savings: 34.411 MJ/day * 0.0078 EUR/MJ natural gas * 0.6 = 122 EUR/day = 45,000 EUR/year.

It should be noted that if total heat energy could be sold, the revenue and savings would reach EUR 203/day and EUR 74,000/year. Th s would be possible if sales were not for the heat-variable demands of the district heating system, but for the suffici tly high constant heat demand of an industrial consumer in a nearby industrial park (e.g. a bioethanol plant or a slaughterhouse) when the heat would be bought at the natural gas price. However, the latter is in practice much more insecure.

In the case of district heat sales, together with the electricity, revenue is 198,000 + 45,000 = 243,000 EUR/year.

The investment cost of CHP technology (with 266–280 kWe capacity) (with existing rotting equipment) following our own calculations is 231,000 EUR.

Biomethane production

For the calculation of biomethane production, the following basic data were used:

- The amount of biogas produced: 1860 Nm^3/day (from the previous calculation)
- The methane content of biogas: 65%
- Cleaning technology: 2% methane loss and 96% methane gas recovery
- The calorific v lue of the biomethane obtained: 34 MJ/Nm^3
- Natural gas purchase price: 0.0078 EUR/MJ (EU wholesale average price)
- CNG/CBM density: 0.78 kg/Nm^3
- CNG average price: 1.18 EUR/kg

The amount of biomethane produced is 1860 Nm^3/day * 65% biogas methane content * 98% effici cy/96% biomethane methane content = 1234 Nm^3/day biomethane.

In the case of supply to the natural gas network, the expected revenue is 1234 Nm^3/day biomethane * 34 MJ/Nm3 * 0.0078 EUR/MJ = 327 EUR/day = 119,000 EUR.

When using the output as a fuel (CNG/CBM), since the prices are expressed in kg, it is therefore appropriate to convert the resulting quantity into this unit of measure: 1234 Nm^3/day biomethane = 963 kg/day biomethane.

Value of this: 963 kg/day biomethane * 1.18 EUR/kg = 1136 EUR/day = 415 EUR/ year. Th s may further contribute to the potential recovery of CO_2 obtained during the cleaning process.

The value of the required cleaning equipment (120–130 Nm^3/h capacity), based on our own calculations: 860,000 EUR (this does not include investment costs related to sales).

However, it is important to take into account that if we convert biogas to biomethane, we need to purchase the sewage plant's self-consumption needs. Their values in the present case are:

Electricity: 7709 kWh/day * 0.112 EUR/kWh = 863 EUR/day

Heat: 18,500 MJ/day * 0.0078 EUR/MJ = 144 EUR/day

Total: 1007 EUR/day = 368,000 EUR/year

The remaining sales revenue from biomethane sales:

Natural gas sales: 119–368 = (-) 249,000 EUR/year

Sales of propellants: 415–368 = 47,000 EUR/year

Effects on CO_2 emission

When calculating the effects of CBM on the CO_2 emission, we should start the calculation with the emission values of the given technological process (CHP, upgrading). Then we need to make a comparison between the CO_2 emissions of the used and of the substituted fuels (electricity, natural gas, diesel fuel). We used the following emission factors in our calculations (**Table 5**).

	Value	Measure	Source
Substitution of fuel			
Diesel	69.95	kg CO_2eq/GJ	[52]
CNG	49.75	kg CO_2eq/GJ	[52]
Electricity	391	kg CO_2eq/kWh	[53]
Operation of sewage plants			
CHP	11	kg CO_2eq/GJ	[37]
Upgrading and feed-in station	41	kg CO_2eq/GJ	[37]
Upgrading and fuelling	22	kg CO_2eq/GJ	[37]

Table 5.
Substitution and operation factors.

The use of the systemic approach in city operation and regional development requires that the interest and demand of the environment, the local communities and the economy would be satisfi d on a mutually benefic al level [54]. We also used this approach.

If the sewage plant itself was evaluated environmentally, the decision would be very simple: because of lack of CBM-fuelled vehicle, the CHP technology would be considered as the most favourable option. At the level of municipality (considering the waste management, local transport and central heating system), we need to make a three-sided comparative analysis.

In the case of the conventional CHP technology, direct emission from the operation should be considered as basis, and it should be reduced with the self-consumed heat and electricity. In this case the saved emission of local transport and of central heating remains the same. Regarding the other two technologies, the emission savings will be derived from the saved natural gas or diesel fuel of buses, respectively (**Table 6**).

Technology	CHP		Upgrading + gas pipeline	Upgrading + transport
	Heat	Electricity	Natural gas	Diesel
	GJ/yr	MWh/yr	GJ/yr	GJ/yr
Savings in fuel (in given unit above)	4357	1815	15,316	15,316
Savings in CO_2eq (t/yr)	217	710	762	1072
Biogas input (GJ/yr)	15,004			
Emission from technology (t/yr)		165	615	330
CO_2eq emission (t/yr)		**-762**	**-147**	**-742**
Source: Own calculations.				

Table 6.
CO_2 emission balance of the evaluated technologies.

Table 6 shows that CBM production for feed-in has the highest energy demand and the lowest emission saving potential. The other two options could be regarded near equal, since though CBM for fuelling has slightly better emission saving potential, it needs more energy for the operation.

Comparison of the technologies examined

The three uses of the biogas produced from the wastewater analysed above can thus be characterised by the following end parameters for a 100,000 PE (population equivalent) sewage capacity (**Table 7**).

Sphere of use	Investment demand (thousand EUR)	Expected revenue (thousand EUR/year)	Emission saving (t CO_2eq/yr)
Cogeneration (electricity and heat)	231	average: 243, max. 272	762
Cleaning (supplying network)	860	-249	147
Cleaning (for fuel)	860	47	742
Source: Own calculations.			

Table 7.
Parameters of the given plant.

The above values show that introducing biomethane into the natural gas network is the least competitive compared to the other two modes of recovery both in economical and in environmental aspects. In the case of existing digesters, the gas engine is very fast—and the most effici t return on investment, if it is connected to a suffici tly large district heating network, ideally an industrial park. Although sales of biomethane for propellants can theoretically achieve the highest turnover on their own, due to the self-consumption needs of the sewage plant, and the investment cost—which is more than three times higher—it is not the most favourable alternative from the sewage plant's perspective. Since the environmental effects are near the same, it can be stated that CHP could be evaluated as the best way for biogas utilisation for the sewage plant.

Economics of the use of CNG at the city level

In general, it can be stated that the long-term spread of a product market (in our case CNG/CBM) is only expected if the product is worthwhile to produce, market and use. In the event of any losses suffered by any actor in the market, in the absence of subsidies, the vertical relationship is interrupted, and so the interest of all actors must be ensured. A subsidy is justifi d by the macroeconomic benefits of the public fi ances (environmental protection, import substitution, employment). In addition, it must, of course, be cheaper and more accessible than competing fuels and must also ensure that the investment needed to operate is recovered within a reasonable period of time. Does biomethane meet all of these criteria under current economic conditions?

It can be clearly seen from the economic data listed in **Table 3** that biomethane can be produced essentially at the consumer price (0.8–1.1 EUR/kg) at relatively small (250 Nm^3/h) biogas plants, while the price-equivalent own costs for its substitutes, i.e. petrol and diesel (0.87–0.98 EUR/l on average), are signifi antly below their average consumer price (1.43–1.44 EUR/l). In the case of non-fi al sales (i.e. not to private individuals) but for business use (buses, machines), equivalent values for petrol and gas oil prices reduced by VAT and other price discounts must be taken into account (cca 1.08 EUR/l in both cases, with an average 25% reduction), with which the smallest agricultural biomethane plants (with a capacity of 250 Nm3/h) are just competitive (0.8–1.1 EUR/kg), while larger farm sizes are able to produce cheaper biomethane. For farm-sized plants, it is economical to produce and use local CBM only when using their own by-products and primarily for the fuel supply of their own vehicles.

Therefore, in the vertical market structure (with the appropriate size and type of user), profits are generated in the current economic conditions, as well. At the same time, in the case of internal use (e.g. operation of a common sewage plant gas and local public transport system), the state loses signifi ant tax revenue, which is, however, easily off et by the externalities present in the public fi ances.

Consequently, the cost of compressed biomethane (CBM) as a propellant is typically just below or at the CNG price level (average 1.18 EUR/kg) but much cheaper than the diesel typically used in local transport vehicles.

Finally, the conversion of vehicles to CNG operation and the return on the CNG premium can be estimated as follows.

In the case of a city bus, taking into account a life expectancy of 1000,000 km, single buses use cca. 370,000 l, and articulated buses cca. 470,000 l of diesel fuel in their lifetime. The fuel cost savings during the CNG operation can be estimated at 37,000 EUR and 47,000 EUR per bus, taking into account the average reclaimed VAT and the wholesale discount. Thus, a local public transport service of county town size (e.g. Debrecen, Hungary, with 210,000 local inhabitants; a fl et 100 of single and 40 articulated buses) can save up to € 5–6 million over the entire lifetime but also hundreds of thousands of EUR on a yearly basis. The expected return on the conversion of a single bus is as follows.

Basic calculation data:

- 60,000 km/year performance (16–17 years calculated lifetime)
- 37 l/100 km gas oil consumption
- An average saving of 0.1 EUR/l of gas oil equivalent with CNG (gas oil wholesale price)
- 40,000 EUR conversion cost/acquisition surplus cost
- Considering 3%/year increase in gas oil price and the same as opportunity cost (assuming use of own money in fi ancing)

Based on CNG purchased on the basis of this data, the investment will be repaid after approximately 18 years without subsidy, so slightly beyond the useful life of the vehicle; however, with a subsidy the payback period is much shorter, and the benefits of environmental protection have not yet been taken into consideration. Increasing performance (better use of buses) and rising gas oil prices can signifi antly reduce the payback period. Since gas fuelling has a positive effect on engine wear, it does not endanger the potential performance; it is rather determined by the transport features. One-hundred thousand kilometre use of bus per year results in 11-year-long repayment period.

In the case of CBM produced in its own wastewater plant, savings are much greater than with diesel; therefore, with the same technical parameters, the payback period will be reduced, even without subsidy.

Environmental advantages of a CNG-/CBG-fuelled bus fl et are also signifi ant. Considering the average value of 18 kg/GJ GHG saving of CNG/CBG compared to gas oil and the bus fl et of Debrecen with the above-mentioned characteristics, the GHG saving can reach 14.4 tonnes/bus in a year, or 15,000–20,000 tonnes/bus fl et, moreover in the city centre, mostly suffering air pollution.

Average data used for calculation of passenger cars:

- 15,000 km/year performance (15-year lifetime)
- 8 l/100 km of petrol consumption

- An average saving of 0.57 EUR/l of petrol equivalent with CNG (at consumer prices)
- Conversion cost of EUR 1600 and extra purchase cost of EUR 4000
- Considering 3%/year increase in petrol price and the same as opportunity cost (assuming use of own money in fi ancing)

In the case of the above parameters, the conversion is expected to take place within 2–3 years, while the purchase price of the new car would be repayable in 5–6 years, mainly depending on the mileage and current petrol prices, if there was no problem with refuelling. In the case of a local public CNG filling station, gasfuelled cars can be recommended primarily to those private individuals who are involved in local transport and travel long distances in a year (e.g. local taxi drivers) or those that are more environmentally sensitive and thus appreciate the benefits of using gas-fuelled cars.

Reference plants

Biomethane production based on various types of waste and its use as a propellant can be found in several places. At this point, we will introduce some international examples, focusing in more detail on wastewater-based biomethane production.

As the study by Barisa et al. [55] shows, there are many potential waste-based raw materials available to a settlement that are suitable for biogas and biomethane production:

- The organic proportion of unsorted municipal solid waste (MSW)
- Separately collected biowaste from restaurants and grocery stores
- Separately collected green waste from garden and park management
- Wastewater sludge (including the amount generated by the dairy plant)
- Organic waste from the industry (e.g. brewer's grain)

With regard to their available volume, it can be said in general that in a given settlement the municipal solid waste, separated green waste and sewage sludge produced in the sewage plant make up the largest amount. However, considering the costs of collecting and separating these three types of raw materials, there may be signifi ant differences. The utilisation of sewage sludge in the sewage plant— continuously and, in a relatively homogeneous amount, free of charge—can be considered cost-effective in this respect. In addition, other waste materials can be used safely for biogas production and its subsequent purifi ation in sewage plants.

In practice, wastewater treatment plants in many cases include organic food waste/by-products that contribute to improving the carbon-nitrogen ratio of sewage and sewage sludge, thereby achieving higher biogas and biomethane yields.

In what follows we present a few examples of biogas or biomethane production plants based on organic waste or sewage.

Sweden

In Linköping, Sweden, biomethane is used in urban transport, not only for buses and heavy and light motor vehicles but also for trains [56]. The total cost of EUR 14,000,000 invested in 1996 can be mentioned as one of the successful examples of the integration of fuel supply for agriculture, the community and individual transportation. In the Linköping waste-to-energy plant, biogas production was initially based on the by-products and wastes of the crop and livestock (slaughterhouse) sector, while in the framework of a development programme, from 2001, they have also produced renewable propellants from organic waste from public institutions and restaurants [57]. Since 2002, there are only biogas buses in the urban transport fl et, and the CO_2 emissions have been reduced by more than 9000 tonnes per year [58].

Another Swedish example is the Nordvästra Skånes Renhållning AB (NSR) biomethane plant in Helsingborg, which generates 80 GWh of biomethane per year from 160,000 tonnes of separated food waste. The methane produced is supplied to the grid and is used for the operation of trucks, taxis and private cars. From 160,000 tonnes of digested food waste in the biogas plant, approximately 490 tonnes of N, 90 tonnes of P and 170 tonnes of K are available for recirculation as fertilizer each year [59].

Another interesting example is the Swedish city of Uppsala. As early as 1996, animal manure and slaughterhouse waste were used for biogas production, and then for biomethane production after purifi ation, which was used for the operation of buses. Thereafter, developments in two stages up to 2010 resulted in the production of biomethane from signifi ant quantities of organic waste from their own city and other settlements; annual production has reached 3000,000 Nm^3 [60]. Overall 71 of the city buses were fuelled by biomethane, which amounts for 35% of fuel used in public transport in Uppsala in 2014 [61].

Considering the Swedish examples, it is not surprising that in Swedish households 60% of organic waste is collected separately and utilised.

Sewage water-based biomethane production was implemented in Hammarby Sjöstad (Stockholm), Sweden. Within the framework of the project, an integrated closed wastewater-energy system has been implemented based on local authority/ municipal sewage. After the sewage is purifi d in the system, propellant biogas and biomethane are also produced, as well as heat and electricity. Hammarby Sjöstad is located in one of the most progressive cities in the world with regard to sustainability. The city has reduced carbon emissions by 25% per resident since 1990 and has established a target of reducing emissions to 3 tonnes of CO_2 per capita in 2015. Th s value is extremely low for developed countries, considering the entire country of Sweden has an average emission rate of 4.5 tonnes of CO_2 per capita, while the average for Europe is approx. 6.5 tonnes per capita, and the average for the United States is 16.5 tonnes per capita [62, 63].

Hungary

Sewage-based, biomethane propellant production was also implemented in Zalaegerszeg (Hungary) (**Figure 3**). The investment began in 2011 and cost 140 million HUF (about 444,000 EUR), of which HUF 120 million was for the biogas

Figure 3.
Zalaegerszeg wastewater treatment plant with biomethane filling station. Source: Own photo.

cleaning system, while the cost of the filling station was HUF 20 million. Daily biomethane production is 3600 m^3, while the biomethane's unit cost is 0.52 EUR/m^3 [64]. Although this value is slightly higher than the consumer price of natural gas, it is just about half as much as the price of diesel oil. Since the substituted buses typically run on diesel oil, the use of CBG should result positive economic effect. The composition of the biogas is 65% methane, 30% carbon dioxide and 5% other gases, including hydrogen sulphide. Raw biogas signifi antly reduces the lifetime of machines because of its hydrogen sulphide content. For a safe and effici t use, a multi-stage cleaning process is initiated: gas is pressurised through an activated carbon filter, which reduces sulphur content, other gas content and humidity. Subsequently, 75% methane and 25% carbon dioxide gas is subjected to an aqueous wash, whereby the carbon dioxide is converted to carbonic acid and transferred from the biogas to the aqueous solution. During the production of biomethane, the purifi d biogas is pressed at a pressure of about 6 bars into the washer, in which the chemical transmission process occurs. After aqueous washing, a 99.7% methane gas is produced, which is excellent for use in CNG vehicles. The biofuel produced fuels for municipal vehicles (12 vans and 3 buses). The CO2 saving can be estimated 22 tonnes/year [65, 66].

United Kingdom

At the beginning of the 2000s, the largest wastewater-based power plant in the United Kingdom was operating at Minworth. During the operation of the plant, several technological developments were introduced, resulting in an increasing amount of biogas production.

The total investment cost of the plant was €8,150,000, which allowed the production of 3400 Nm^3/h of biogas (nearly half of which is upgraded to biomethane) with 16 digesters, totalling 80,000 m^3/day. The type of raw material is sewage sludge (4000 m^3/h). The development of biomethanic purifi ation is due to two important factors: (1) simultaneously with the increase in the quantity of biogas produced—due to the variable amount of energy—the amount of electricity that can be supplied to the electricity grid was limited; (2) a signifi ant amount (40%) of heat loss was also a serious problem. All of this led to the construction of biomethane purifi ation capacity, which keeps the energy produced from sewage well stored and also avoids heat loss. The unit uses absorption, which provides a high-pressure water scrubbing process to clean biogas and to increase the methane concentration [67]. With an annual biomethane production of 63,000,000 kWh, Minworth is the fi st connection of unconventional gas into the local transmission system, with a signifi ant impact on reducing CO_2 emissions [68].

Other examples of good practice

In the following we present other international examples, and cases of good practice of use as a fuel and for supply to the natural gas grid, using as a source the Bin2GRID (2016) project [69]:

Good practice for usage of biomethane and CNG vehicles:

- Margarethen am Moos (Austria)
- Rechnitz (Austria)
- Norrköping (Sweden)
- St. Gallen (Switzerland)
- Lille (France)
- Madrid (Spain)
- Sevilla (Spain)

Good practice of biomethane injection into the natural gas grid:

- Margarethen am Moos (Austria)

- Vienna (Austria)
- Utzenstorf (Switzerland)
- Norrköping (Sweden)
- Madrid (Spain)

Conclusions

Public transport, sewage treatment and district heating are mostly operated by local authorities and can therefore be influenced by local decisions, so biomethane can be well integrated into local transport systems, and the terms of sale are not dependent on gas suppliers, either. The latter may be signifi ant in the event of potential future gas price rises, which does not affect the cost of locally produced biomethane. At the same time, it should be noted that the economic interests of biomethane-producing wastewater plants are rather geared towards traditional cogeneration technologies rather than biomethane production because of the signifi ant electricity and heat demand of sewage treatment technologies.

From an environmental point of view, it is also noteworthy that the reduction of the emissions of cadmium due to the use of local public transport with CNG occurs right in the city centre, where air pollution is the most serious problem. From the operation point of view, it is essential that in the case of local transport, there is no danger of emptying fuel tanks, since the filling station is available locally, unlike with long-distance transport. The public education objective of technology is not negligible as many people (those travelling by public transport) are affected by the use of environmentally friendly fuel.

Midterm proliferation is expected to take place, given that in October 2014 the European Parliament and the Commission adopted the "Clean Fuel for Transport" package, which obliges member states to take the fi st steps by 2020 in cities and suburbs and by 2025 on motorways, to make CNG fuelling stations for cars and then allow transport by LNG trucks on Europe's main transport lanes.

Due to the above, the production of biomethane at metropolitan wastewater plants for transport purposes can serve the interests of residents in an effici t way, both economically and environmentally, also enabling them to enjoy a kind of energy self-suffici cy.

Acknowledgements

The publication was supported by EFOP 3.6.3-VEKOP-16-2017-00007 – "Young researchers for talent"– Supporting careers in research activities in higher education.

Conflict of interest

The authors declare no confli t of interest.

Author details

Attila Bai* and Zoltán Gabnai
Faculty of Economics and Business, University of Debrecen, Debrecen, Hungary

*Address all correspondence to: bai.attila@econ.unideb.hu

References

[1] Eurostat. Share of Transport Fuel From Renewable Energy Sources [Internet]. 2018. Available from: https:// ec.europa.e u/eurostat/web/products-eurostat-ne ws/-/ DDN-20180312-1?inheritRedirect= true [Accessed: October 28, 2018]

[2] Pintér G, Baranyai NH, Wiliams A, Zsiborács H. Study of photovoltaics and LED energy effici cy: Case study in Hungary. Energies. 2018;11(4):790. DOI: 10.3390/en11040790

[3] Fraunhofer ISE. Levelized Cost of Electricity—Renewable Energy Technologies [Internet]. 2018. Fraunhofer Institute for Solar Energy Systems ISE. Available from: https:// www.ise.fraunhofer.de/content/dam/ ise/en/documents/publications/studies/ EN2018_Fraunhofer-ISE_LCOE_Rene wable_Energy_Technologies.pdf [Accessed: October 15, 2018]

[4] European Commission. Legal Sources on Renewable Energy [Internet]. 2018. Available from: http:// www.res-legal.eu/search-by-country [Accessed: October 10, 2018]

[5] Szunyog I. Quality requirements of the application of biogases in natural gas public utility services in Hungary [Thesis]. Miskolc: University of Miskolc; 2009

[6] Bai A, Jobbágy P, Popp J, Farkas F, Grasselli G, Szendrei J, et al. Technical and environmental effects of biodiesel use in local public transport. Transportation Research, Part D: Transport and Environment. 2016;47: 323-335. DOI: 10.1016/j.trd.2016.06.009

[7] Hakawati R, Smyth BM, McCullough G, De Rosa F, Rooney D. What is the most energy effici t route for biogas utilization: Heat, electricity or transport? Applied Energy. 2017;206: 1076-1087. DOI: 10.1016/j. apenergy.2017.08.068

[8] Jámbor A, Mizik T. Bioethanol—Who is the winner? In: Schäfer C, Rupschus C, Nagel UJ, editors. Enhancing the Capacities of Agricultural Systems and Producers. Weikersheim: MACE, Margraf Publishers; 2008. pp. 210-215

[9] Santos G. Road transport and CO2 emissions: What are the challenges? Transport Policy. 2017;59:71-74. DOI: 10.1016/j.tranpol.2017.06.007

[10] Cullinane S, Cullinane K. Car dependence in a public transport dominated city: Evidence from Hong Kong. Transportation Research, Part D: Transport and Environment. 2003;8(2): 129-138. DOI: 10.1016/S1361-9209(02) 00037-8

[11] NGVJournal. Worldwide NGV Statistics [Internet]. 2018. Available from: http://www.ngvjournal.com/? page_id=22218 [Accessed: September 19, 2018]

[12] Natural and Biogas Vehicle Association (NGVA). Statistical Report 2017 [Internet]. The Natural & Bio Gas Vehicle Association. 2017. Available from: https://www.ngva.eu/wp-conte nt/uploads/2018/01/170648_NGVA_ Europe_statistical-Report_2017_5-2.pdf [Accessed: September 05, 2018]

[13] European Biogas Association. EBA Statistical Report 2017 [Internet]. Available from: http://european-biogas.eu/2017/12/14/eba-statistical-report-2017-published-soon/ [Accessed: August 08, 2018]

[14] Bagi Z, Ács N, Bálint B, Horváth L, Dobó K, Perei KR, et al. Biotechnological intensifi ation of biogas production. Applied Microbiology and Biotechnology. 2007; 76(2):473-482

[15] Scarlat N, Dallemand JF, Fahl F. Biogas: Developments and perspectives in Europe. Renewable Energy. 2018;129: 457-472. DOI: 10.1016/j.renene.2018. 03.006

[16] Eurostat. European Statistics 2017 [Internet]. Available from: http://ec.e uropa.eu/eurostat [Accessed: June 05, 2018]

[17] Scandinavian Biogas. Renewable Fuels are Strategically Important [Internet]. 2018. Available from: http:// scandinavian-biogas.com/en/about-b iogas/ [Accessed: November 02, 2018]

[18] Yang L, Ge X, Wan C, Yu F, Li Y. Progress and perspectives in converting biogas to transportation fuels. Renewable and Sustainable Energy Reviews. 2014;**40**:1133-1152. DOI: 10.1016/j.rser.2014.08.008

[19] Patterson T, Esteves S, Dinsdale R, Guwy A. An evaluation of the policy and techno-economic factors affecting the potential for biogas upgrading for transport fuel use in the UK. Energy Policy. 2011;**39**(3):1806-1816. DOI: 10.1016/j.enpol.2011.01.017

[20] Beil M, Beyrich W. Biogas upgrading into biomethane. In: Wellinger A, Murphy J, Baxter D, editors. Woodhead Publishing Series in Energy. The Biogas Handbook. Cambridge: Woodhead Publishing; 2013. pp. 342-377. ISBN: 9780857094988

[21] Popp J, Pető K, Nagy J. Pesticide productivity and food security. A review. Agronomy for Sustainable Development. 2013;33(1):243-255. DOI: 10.1007/s13593-012-0105-x

[22] Popp J, Kot S, Lakner Z, Oláh J. Biofuel use: Peculiarities and implications. Journal of Security and Sustainability Issues. 2018;7(3):477-494. DOI: 10.9770/jssi.2018.7.3(9)

[23] European Biogas Association. EBA's BIOMETHANE Fact Sheet [Internet]. Available from: http://european-biogas.eu/wp-content/uploads/files/2013/10/ eba_biomethane_factsheet.pdf [Accessed: May 18, 2018]

[24] Domanovszky H, Szalkai I. Energy Aspects of Methane Gas [Internet]. Hungarian Gas-Powered Transport Cluster Association. 2012. pp. 1-7. Available from: www.mgkke.hu (In Hungarian) [Accessed: October 20, 2017]

[25] Åhman M. Biomethane in the transport sector—An appraisal of the forgotten option. Energy Policy. 2010; 38(1):208-217. DOI: 10.1016/j. enpol.2009.09.007

[26] Bordelanne O, Montero M, Bravin F, Prieur-Vernat A, Oliveti-Selmi O, Pierre H, et al. Biomethane CNG hybrid: A reduction by more than 80% of the greenhouse gases emissions compared to gasoline. Journal of Natural Gas Science and Engineering. 2011;3(5):617-624

[27] Fuchsz M, Kohlheb N. Comparison of the environmental effects of manure- and crop-based agricultural biogas plants using life cycle analysis. Journal of Cleaner Production. 2015;86:60-66. DOI: 10.1016/j.jclepro.2014.08.058

[28] Ryan F, Caulfi ld B. Examining the benefits of using bio-CNG in urban bus operations. Transportation Research, Part D: Transport and Environment. 2010;15(6):362-365. DOI: 10.1016/j. trd.2010.04.002

[29] European Environment Agency. TRACCS Database [Internet]. 2013. Available from: https://www.eea.europa. eu/data-and-maps/data/external/traccs [Accessed: November 03, 2018]

[30] Transport Science Institute [Internet]. 2014. Available from: http:// www.kti.hu/trend/fenntarthato-kornye zet/page/4/ [Accessed: October 08, 2018]

[31] Lim C, Kim D, Song C, Kim J, Han J, Cha JS. Performance and emission characteristics of a vehicle fueled with enriched biogas and natural gases. Applied Energy. 2015;139:17-29. DOI: 10.1016/j.apenergy.2014.10.084

[32] Subramanian KA, Mathad VC, Vijay VK, Subbarao PMV. Comparative evaluation of emission and fuel economy of an automotive spark ignition vehicle fuelled with methane enriched biogas and CNG using chassis dynamometer. Applied Energy. 2013; 105:17-29. DOI: 10.1016/j. apenergy.2012.12.011

[33] Directive 2009/30/EC of the European Parliament and of the Council of 23 April 2009 amending Directive 98/ 70/ EC as regards the specifi ation of petrol, diesel and gas-oil and introducing a mechanism to monitor and reduce greenhouse gas emissions and amending Council Directive 1999/32/EC as regards the specifi ation of fuel used by inland waterway vessels and repealing Directive 93/12/EEC. Available from: https://eur-lex.europa.eu/legal-content/EN/TXT/?uri=celex%3A32009L0030 [Accessed: May 28, 2019]

[34] Goulding D, Power N. Which is the preferable biogas utilisation technology for anaerobic digestion of agricultural crops in Ireland: Biogas to CHP or biomethane as a transport fuel? Renewable Energy. 2013;53:121-131. DOI: 10.1016/j. renene.2012.11.001

[35] Marsden G, Rye T. The governance of transport and climate change. Journal of Transport Geography. 2010;18(6): 669-678. DOI: 10.1016/j. jtrangeo.2009.09.014

[36] Domanovszky H. Methane Gas in Transport [Internet]. Hungarian Gas- Powered Transport Cluster Association. 2011. pp. 1-8. Available from: www.mg kke.hu (In Hungarian) [Accessed: October 22, 2017]

[37] Patrizio P, Leduc S, Chinese D, Dotzauer E, Kraxner F. Biomethane as transport fuel—A comparison with other biogas utilization pathways in northern Italy. Applied Energy. 2015; 157:25-34. DOI: 10.1016/j. apenergy.2015.07.074

[38] Jobbágy P, Bai A, Juhász IL. Biomethane's Perspectives in Domestic Traffic Study. Agricultural Technology. Sustainable Bioenergy Production [Mezőgazdasági Technika. Fentartható Bioenergia-Termelés Különszám]. Gödöllő: FVM Műszaki Intézet; 2010. pp. 16-18. ISBN: 978-963-611-456-5 (In Hungarian)

[39] Bai A, Stündl L, Bársony P, Jobbágy P, Herpergel Z, Fehér M, et al. Algae production on pig sludge. Agronomy for Sustainable Development. 2012;32(3): 611-618. DOI: 10.1007/s13593-011-0077-2

[40] Bai A, Popp J, Pető K, Szőke I, Harangi-Rákos M, Gabnai Z. The signifi ance of forests and algae in CO2 balance: A Hungarian case study. Sustainability. 2017;9:857-880. DOI: 10.3390/su9050857

[41] Vityi A, Marosvölgyi B. New tree species for agroforestry and energy purposes. In: Proceedings of the 2014 International Conference on Biology and Biomedicine II (BIO '14); 2-4 April 2014; Prague, Czech Republic; 2014. pp. 82-84. ISBN: 978-1-61804-232-3

[42] Kurucz E, Antal G, Fári MG, Popp J. Cost-effective mass propagation of virginia fanpetals (*Sida hermaphrodita* (L.) Rusby) from seeds. Environmental Engineering and Management Journal. 2014;13:2845-2852

[43] Farkas F, Nagy V, Bai A. Tests of differential diesel fuels in engine testing room. APSTRACT—Applied Studies in Agribusiness and Commerce. 2014;8(1): 59-67. ISSN: 1789-7874

[44] CNGEurope. Map of Natural Gas Vehicle (NVG) Compressed Natural Gas (CNG) Filling Stations in Europe [Internet]. 2018. Available from: http:// cngeurope.com/ [Accessed: November 08, 2018]

[45] Autotraveler. Fuel Price in Europe 2018 [Internet]. 2018. Available from: https://autotraveler.ru/en/spravka/fuel- price-in-europe.html#.W_Hj_zhKgdV [Accessed: October 18, 2018]

[46] CNGPort [Internet]. 2018. Available from: http://www.cngport.hu/ [Accessed: October 30, 2018]

[47] Oel-alle [Internet]. 2018. Available from: http://www.oel-alle.de/ [Accessed: October 14, 2018]

[48] Biokraft toff [Internet]. 2013. Available from: http://www.bio-kraf- stoff.de/ [Accessed: March 05, 2014]

[49] European Commission. Electricity Prices for Household Consumers [Internet]. 2018. Available from: https:// ec.europa.eu/eurostat/statistics-expla ined/index.php/Electricity_price_sta tistics#Electricity_prices_for_househ old_consumers [Accessed: November 04, 2018]

[50] Eurostat. Natural Gas Price Statistics [Internet]. 2018. Available from: https://ec.europa.eu/eurostat/sta tistics-explained/index.php/Natural_ga s_price_statistics [Accessed: October 16, 2018]

[51] Bányai Z, Thury P, Kárpáti Á. Energiahatékonyság a Szennyvíztisztításnál. 2014. pp. 96-104. Available from: http://vizugy.uni-pa nnon.hu/content/administration/tana nyagok/Eleveniszapos_rendszerek.pdf [Accessed: May 10, 2018]

[52] Kárpáti Á. Wastewater treatment— Energy management in public wastewater treatment plants. MASZESZ Hírcsatorna. 2016;3:6-20. (In Hungarian)

[53] Joint Research Centre of the European Commission. CoM Default Emission Factors for the Member States of the European Union. Dataset Version 2017 [Internet]. Available from: http:// data.jrc.ec.europa.eu/dataset/jrc- com-ef-comw-ef-2017 [Accessed: September 28, 2018]

[54] Kiss T. Systemic approach in city operation—Possible synergies between waste management systems and energetics. Pollack Periodica. 2012;7(3): 55-63

[55] Barisa A, Dzene I, Rosa M, Dobraja K. Waste-to-biomethane concept application: A case study of Valmiera city in Latvia. Environmental and Climate Technologies. 2015;15(1):48-58

[56] Makareviciene V, Sendzikiene E, Pukalskas S, Rimkus A, Vegneris R. Performance and emission characteristics of biogas used in diesel engine operation. Energy Conversion and Management. 2013;75:224-233. DOI: 10.1016/j.enconman.2013.06.012

[57] GIZ and ICLEI. Linköping, Sweden. Waste-to-Energy Power Plant: Biogas Powers Public Transport in Linköping [Internet]. 2014. Available from: http:// www2.giz.de/wbf/4tDx9kw63gma/14_ UrbanNEXUS_CaseStory_Linkoping. pdf [Accessed: September 15, 2018]

[58] IEA Bioenergy Task 37. 100% Biogas for Urban Transport in Linköping, Sweden. Biogas in Buses, Cars and Trains [Internet]. Available from: http:// www.iea-biogas.net/files/daten-reda ktion/download/linkoping_fi al.pdf [Accessed: May 06, 2019]

[59] IEA Bioenergy Task 37. More than 10 Years Production of Fossil Free Automotive Fuel and Certifi d Digestate from Food Waste. Vera Park in Helsingborg, Sweden [Internet]. 2014. Available from: http://www.iea-biogas. net/files/daten-redaktion/download/Success%20Stories/NSR_Sweden_web. pdf [Accessed: October 23, 2018]

[60] European Biogas Association. Success Stories: Anaerobic Digestion of Biodegradable Municipal Solid Waste in European Cities [Internet]. Available from: http://european-biogas.eu/wp- content/uploads/2016/03/Urban-Biowa ste-AD-Success-Stories-FINAL.pdf [Accessed: October 29, 2018]

[61] UNIZAG FSB. Factsheets on Good Practice of Biogas Upgrade. Project 'BIN2GRID' [Internet]. Available from: http://www.bin2grid.eu/documents/ 73603/136970/Eng_Bin2Grid_revision. pdf/2dbe8c8b-1656-4336-8438-a15fcd632331 [Accessed: May 06, 2019]

[62] Foletta N. Hammarby Sjöstad. Stockholm, Sweden. Case Study. ITDP Europe [Internet]. 2014. Available from: https://www.itdp.org/wp-content/ uploads/2014/07/20.-092211_ITDP_ NED_Hammarby.pdf [Accessed: May 06, 2019]

[63] The World Bank. CO2 Emissions (Metric Tons per Capita). 2019. Available from: https://data.worldbank. org/indicator/en.atm.co2e.pc [Accessed: May 06, 2019]

[64] PAN-LNG Project. PAN-LNG Project Study 1.7. [Internet]. 189p. Available from: http://www.panlng.e u/wp-content/uploads/2016/06/1_7_ PAN-LNG_biogaz.pdf. [Accessed: August 17, 2018]

[65] Intelligent Energy Europe Bio- Methane Regions. Promotion of Bio- Methane and Its Market Development Th ough Local and Regional Partnerships [Internet]. Available from: https://www.ea-stmk.at/documents/ 20181/25550/D_5_1_Best_Practice_ Monitoring.pdf/13b61dbe-a886-454d-97e4-24aeada73cab [Accessed: October 24, 2018]

[66] CNGPort [Internet]. 2018. Available from: http://www.cngport.hu/hire k/munkaba-allt-zalaegerszeg-elso-cng-s-busza.html [Accessed: October 30, 2018]

[67] European Biogas Association. Biomethane from Sewage Sludge as a Successful Business Model in Minworth, UK [Internet]. Available from: h ttp://european-biogas.eu/wp-content/ uploads/2015/03/Success-story-Malmbe rg-UK.pdf [Accessed: October 27, 2018]

[68] Renewable Energy Association. UK Biomethane. Market Report [Internet]. 2016. Available from: https://www.r-e-a.net/upload/rea_uk_biomethane_market_ report_11-04-16small.pdf [Accessed: May 06, 2019]

[69] Güssing Energy Technologies GmbH. Project 'BIN2GRID'Report on Good Practice of Biomethane Usage as a Transportation Fuel [Internet]. Available from: http://www.bin2grid.e u/documents/73603/136990/ D6.1-Good-practise-biomethane-tra nsport-fuel_c.pdf/ [Accessed: November 02, 2018]

2

Integrated Life Cycle Economic and Environmental Impact Assessment for Transportation Infrastructure

Jiawen Liu, Hui Li, Yu Wang and Nailing Ge

Abstract

In order to realize the sustainable development of transportation infrastructure, more and more attention has been paid to the multi-scheme selection method of road engineering, while the existing life cycle cost analysis (LCCA) and life cycle assessment (LCA) methods are often isolated from each other, which cannot better realize the comprehensive evaluation of road life cycle. Th s chapter will review and summarize the development of LCCA and LCA systematically. Pointing out the existing problems in current research, the idea of integrated evaluation method combining LCA and LCCA is proposed. It puts forward the future development direction based on the deficie cy of the current research results and provides useful reference for the popularization and application of the life cycle methods in road engineering.

Keywords: life cycle assessment, life cycle cost analysis, transportation infrastructure

Introduction

According to the defin tion of ISO 14040, life cycle refers to the continuous and interrelated stage of the product system, which generally starts from the acquisition of raw materials or products from natural resources and ends with the fi al treatment. It considers the planning, design, production, distribution, operation, use, maintenance, and recycling of the product, from the initial or design phase of the product, as shown in **Figure 1.**

Existing life cycle methods include life cycle cost analysis (LCCA) and life cycle assessment (LCA). Similar to the engineering budget method, the LCCA extends the time range of evaluation to the whole life of the product and focuses on its use, maintenance, and recycling, making the evaluation of the product more comprehensive and reasonable. With the increasing awareness of environmental protection, LCA, a method of evaluating product life cycle impact from the perspective of environmental impact rather than economic cost, has also been further developed. Now the application of LCA has become mature in many fi lds. In the early twenty-fi st century, the international road engineering research began to introduce LCA and developed a series of professional LCA analysis software. ISO 14040 points out that LCA method is composed of four parts: defin tion of goals and scope, inventory analysis, impact assessment, and result interpretation. It studies and analyzes the stages of raw material acquisition, construction, use, maintenance, and end of life, which is of great signifi ance for promoting the ecological development of road construction. However, existing analysis isolates economic costs and environmental impacts from each other and fails to fully explore the overall impact of the product.

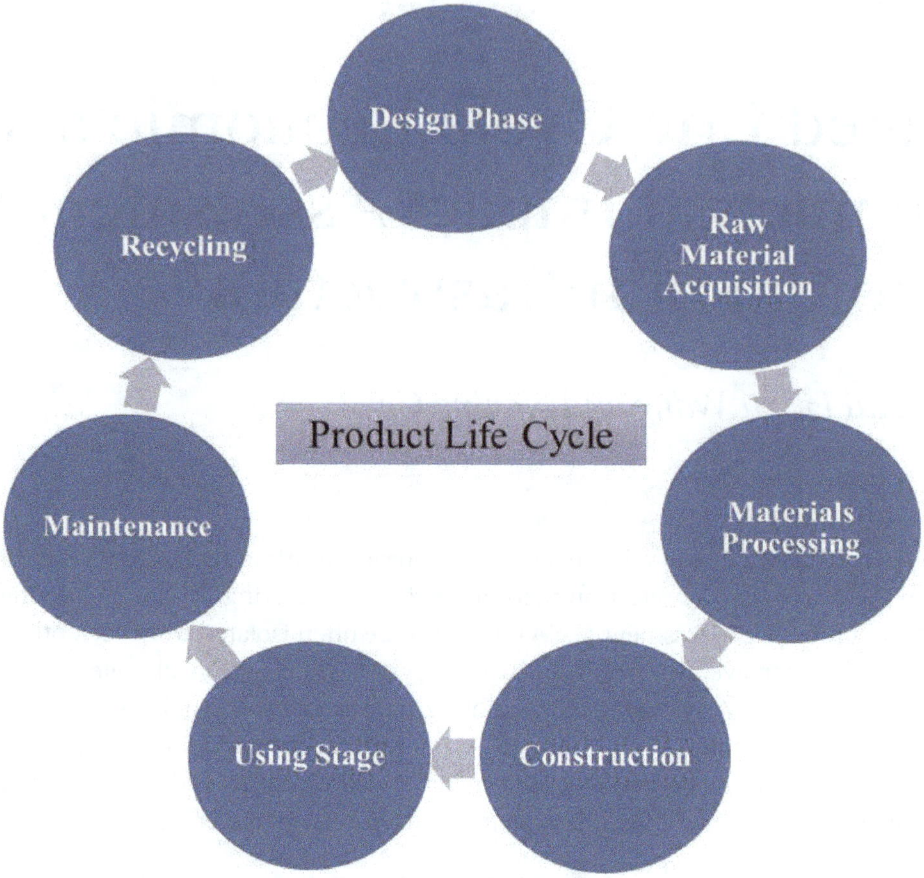

Figure 1.
Product life cycle.

Th s chapter will summarize the international research development of LCCA and LCA applied on transportation infrastructure and puts forward the idea of evaluating the whole life cycle by combining the two life cycle methods; the calculation models involved in the integrated method will also be introduced, so as to provide reference for the decision of multi-scheme comparison in road engineering and the popularization and application of the life cycle methods.

Development of LCCA and LCA

Development of life cycle cost analysis

The concept of life cycle cost analysis (LCCA) has been introduced into the road engineering since the 1960s, when it was proposed by the US military and applied to the procurement of military equipment, aiming to solve the problem of deterioration of pavement performance and increase of maintenance costs at that time. The AASHTO pavement design guide of 1986, 1993, and 2002 all required the use of LCCA for the comparison and selection of the scheme, and the design specifi ations prepared by the National Highway Network in 1995 clearly stipulated that a complete LCCA report must be made for government investment projects exceeding \$25 million [1].

Walls and Smith [31], issued a technical bulletin on LCCA, introducing a detailed method for calculating user costs in the operating area and introducing probabilistic methods to discuss the uncertainty of LCCA.

The announcement fi st discusses the basic principles widely followed by LCCA and then gives the application case of traditional LCCA pavement design. Secondly, the uncertainty of input parameters is discussed, and the acceptable range of time and discount rate is provided. Thi dly, the sensitivity analysis of traditional LCCA method is discussed. Finally, it proposes the specific contents of user cost including delay cost, vehicle operation cost, and accident cost and presents the specific alculation method [2].

Delwar and Papagiannakis evaluated government and user costs of roads under a variety of road and traffi conditions, based on data from the Department of Transport in Washington. The evaluation results show that the user cost may be signifi antly higher than the government cost, so the user cost cannot be ignored in the life cycle cost analysis [3].

Chan et al. studied and analyzed the accuracy of LCCA decision-making and the accuracy of life cycle costs based on data and case studies from the Michigan Department of Transportation. The results show that LCCA can correctly predict and select the lower cost pavement scheme, but the actual cost is usually lower than the estimated value of LCCA. Th s result may be due to the inadequate consideration of specific pavement characteristics in cost estimation, so improving the process of pavement construction and maintenance cost estimation can help to realize the potential of LCCA pavement scheme selection [4].

To sum up, after 10 years of development, the analysis method of LCCA is relatively mature at present, but there is still a lack of data needed for evaluation. Since its inception, LCCA has been widely used in the road industry and has become a necessary component of road program evaluation in the United States. So far, the classifi ation and calculation methods of owner cost and user cost of LCCA have been relatively mature, and the corresponding calculation tools have been widely used in many states of the United States [5].

Development of life cycle assessment

In the 1970s, the oil crisis caused widespread global interest in energy, and then the world began to see a boom in building energy consumption research. The energy consumption survey of buildings fi st emerged in the United States and the United Kingdom. They mainly inspect the energy consumption of existing buildings, tap their energy saving potential, and carry out energy saving transformation, which is called energy auditing [6]. Initially, researchers abroad concentrated on civil buildings and then gradually extend to all aspects of infrastructure construction. Research on road energy consumption has also appeared relatively early, and a lot of research achievements have been made and applied in practice.

Häkkinen and Mäke lä studied the life cycle of pavement in Finland based on the life cycle assessment theory. Th ough the analysis and comparison of common concrete pavement and stone matrix asphalt (SMA) pavement, the author thinks that in terms of energy consumption, if feedstock energy (refers to the combustion energy contained in raw materials of road construction, which can no longer be used as energy) is taken into consideration, asphalt pavement consumes twice as much energy as cement concrete pavement. If feedstock energy is not included, the energy consumption of these two pavements is equal. In terms of carbon dioxide emissions, common concrete pavement discharges 40–60% more than asphalt pavement, and the difference varies depending on the specific aintenance scheme [7].

Horvath and Hendrickson evaluated hot-mixed asphalt concrete pavement and continuously reinforced concrete pavement (CRCP) in the United States. After analysis and comparison, the author came to the conclusion that during the production stage of materials, the energy consumption of asphalt concrete pavement is about 40% more than that of CRCP, but most of the environmental indicators of asphalt concrete pavement are better than that of CRCP [8].

Roudebush compared and analyzed the cement concrete pavement and asphalt concrete pavement in the United States. The author concluded that the value of the asphalt pavement is approximately one time more than that of the cement pavement, about 90.8%. In the stage of material production and pavement maintenance, the energy value of asphalt concrete is approximately two times that of cement concrete [9]. Berthiaume and Bouchard applied exergy, an energy derivative, to study the energy consumption and environmental impact of asphalt and cement concrete pavement structure in Canada. Exergy describes the energy differences of thermodynamic equilibrium between products, which is a tool for measuring product energy and explaining energy quality differences [10].

Mroueh et al. got rid of the traditional comparison between asphalt and concrete and focused on the evaluation and analysis of the application of industrial by- products in pavement structure. The report analyzed the environmental impact of seven pavement structures with fly ash, crushed concrete waste, and blast furnace slag as the substitutes of original materials [11].

Stripple made a comparatively comprehensive study and comparison between cement concrete pavement and cold mix and hot mix asphalt concrete pavement, including accessory facilities of highway such as vegetation, fence, sign, and so on. According to the report, the energy consumption of cement concrete pavement is higher than that of asphalt concrete pavement. For the asphalt mixtures, they both produce the same amount of energy in the stage of production, but the cold mix one increases the energy consumption due to the addition of emulsifier [12]

Nisbet et al. listed life cycle inventory (LCI) of urban roads and highways in the United States and analyzed the energy consumption of cement concrete pavement and asphalt concrete pavement, respectively. For asphalt concrete pavement, the impact of transportation factors is not obvious. When the feedstock energy of asphalt is included, cement concrete pavement requires less materials, has lower energy resources, and has less exhaust emissions, no matter for urban roads or highways [13].

Park et al. based on the method of composite life cycle assessment, combined with the Korean economy and national energy balance sheet, applied the input– output model to assess energy consumption and gas emissions from roads in material selection and production stages [14].

Zapata and Gambatese [15] found that the results of Horvath and Hendrickson [8] were contrary to those of Stripple [12]. Therefore, by using the same preset conditions as Horvath and Hendrickson, the energy consumption of asphalt concrete pavement and continuous reinforced concrete pavement (CRCP) during the material production and construction stage is analyzed to make a relatively fair comparison. The results show that CRCP consumes more energy in the material production and construction stage, of which the energy consumption of cement production is the main factor, while the drying energy consumption of mixed aggregates is the signifi ant factor affecting the energy consumption of asphalt pavement [15].

The framework of life cycle environmental assessment is relatively complete after years of research, but there are still a lot of deficie cies in the detailed model, and the data collection is also in the initial stage. The framework and theory of LCA have been accurately described in ISO 14040/ISO 14044 series standards, but there are still many different opinions and methods in its application on road.

Application of life cycle assessment in China

In China, research on road energy consumption is mainly carried out from a single aspect, such as production of raw materials or construction technics, but few research focus on the energy consumption during the lifetime of pavement.

From the perspective of economy and energy consumption, Fusen Fang studied and analyzed the cement pavement and asphalt pavement with a life span of 30 years in 1984. The author believes that when the discount rate is no more than 12%, the present value cost of cement concrete pavement is always less than asphalt concrete pavement. However, when the energy contained in asphalt itself is ignored, the energy consumption of cement concrete pavement is 8–17% more than that of the asphalt one [16].

Gu found that the less smooth the pavement, the higher the fuel consumption of the car. By studying the relationship between road surface smoothness and auto-mobile fuel consumption, the author mainly discussed how to improve the pavement smoothness as a way to save energy and gain economic benefits [17].

Ye studied the fatigue and energy consumption optimization design of cementstabilized base. The multilayer elastic system theory is used to directly calculate the stress and strain of pavement structure to obtain its mechanical and fatigue characteristics. Based on this, the paper performs thickness optimization and simple energy consumption analysis and calculation and discusses the technical, economic, and social benefits of cement-stabilized base [18].

Zhang started with the application effect of the old asphalt regenerator researched and produced in Guizhou province and collected relevant work effici cy quota data in the regenerated asphalt pavement project in Anshun, Duyun, and Zunyi. After comprehensive analysis and comparison, it is found that the energy consumption is different due to different seasons [19].

Han made an economic comparison between cement pavement and asphalt pavement in terms of construction cost and fuel consumption, mainly comparing the price of raw materials and the cost of maintenance, and thought that cement pavement has great advantages over asphalt pavement in economy. Moreover, from the perspective of pavement operation, the author analyzed that the fuel consumption of asphalt pavement is about 10% more than cement pavement due to the phenomenon of "defl ction basin" of flex ble structure of asphalt pavement [20].

Yi et al. compiled the energy consumption calculation and environmental assessment methods for the warm mix asphalt (WMA) and half-warm mix asphalt mixture. The analysis shows that the heating of coarse aggregate and the evaporation of water consume nearly 70% energy in the process of mixing. The production temperature has a great influence on the energy loss in the process of asphalt mixture mixing. The higher the production temperature, the more the energy loss. The energy loss during the mixing process of half-warm mixed asphalt mixture is nearly 50% less than that of hot mix asphalt mixture [21].

Shang et al. used the LCA theory and method to divide the life cycle of highways into four stages: material production, construction, maintenance, and dismantling, so as to study the energy consumption and atmospheric emissions within the life cycle of highways. According to the research, the proportion of energy consumption in the production stage of building materials is about 55.7% of the total energy consumption, followed by the maintenance and repair stage 40.5%, the construction stage 5.6%, and the dismantling stage 4%. The results show that most of the highway life cycle energy consumption is the direct and indirect energy consumption in the material production process [22].

Tang and Song summarize the low-energy warm mix asphalt concrete construction technology in the application of G109 national highway rebuilding project and, through the experiment monitoring, found that after the application of the technology, mixing, transporting, and paving of the mixture temperature were signifi antly reduced, saving energy consumption, reducing the CO2 and smoke emissions, and effectively reducing the negative impact of the project which brings to the plateau fragile ecological environment [23]. Pan studied the life cycle energy consumption and carbon emission of highway. After modeling and quantitative analysis of energy consumption and greenhouse gas emissions in different stages of highway life cycle, the author thinks that the average annual energy consumption of cement concrete pavement is less than that of asphalt concrete pavement. Therefore, from the perspective of energy saving, cement concrete pavement is superior to asphalt concrete pavement [24]. Shi analyzed the energy consumption of asphalt pavement regeneration materials in road maintenance. After investigating the energy consumption of five different pavement materials in the four stages of raw material production, mixing plant, transportation, and construction machinery, it was found that the energy consumption of regenerative mixture is less. Therefore, from the perspective of energy saving, the regenerative technology is worth spreading [25].

Ma et al. evaluated the energy consumption of continuous reinforced concrete pavement and hot mix asphalt pavement in the construction process and made it clear that reducing the amount of early energy consumption in the production stage of raw materials in the pavement life cycle is conducive to promoting the sustainable development of the highway engineering fi ld [26].

Li used LCA method to compare the environmental impact of continuous reinforced concrete pavement with asphalt pavement. The energy consumption and emission are quantitatively analyzed by selecting a reasonable calculation method for each stage. Among them, the calculation method of energy consumption includes quota method and IRI—speed—fuel consumption model. The results show that the green degree of continuous reinforced concrete pavement is higher than that of asphalt concrete pavement [27].

Zhang et al. analyzed the influence of different asphalt structural layer design parameters on the carbon emission characterization results. The results show that the greenhouse effect is the most serious in the construction period of asphalt pavement, accounting for more than 95%, and the carbon emissions in the production stage have the greatest impact on the greenhouse effect [28].

Due to the complexity of pavement system, problems still exist in the application of LCA in China:

- There are many differences in the assumptions of system boundary and boundary conditions.
- In the process of inventory analysis, the life stages considered in many studies are not comprehensive.
- Models and methods of life cycle assessment are not unifi d, lacking of consistent criteria for data analysis during the interpretation phase.
- Due to the opacity of domestic industry data, when the LCA method is applied in China, most of the list data are mostly directly from foreign literature or database, so its reliability is difficult to be guaranteed.

Existing problems in current research

There are relatively few studies on the integrated evaluation methods of life cycle economic cost and environmental impact. For the research in this fi ld, the economic cost and environmental impact of multiple schemes are usually calculated separately, and then the advantages and disadvantages of the schemes are compared by multi-objective optimization. Shu et al. fi stly analyzed the differences between the two life cycle methods and introduced the basic principles and disadvantages of PTLaser and TCAce, two software platforms that integrate the two in foreign countries, so as to provide references for domestic researchers [29]. Batouli et al. evaluated the life cycle costs and environmental impacts of different pavement design schemes, and the results showed that the initial cost of flex ble pavement was lower, but it would bring higher long-term costs and environmental impacts [30]. Umer et al. proposed a road scheme evaluation system, integrating LCCA and LCA, and carried out a multi-objective analysis based on economic cost and environmental impact. It is proved that geosynthetics can be used to improve the service life of low-traffi road surface and to minimize the cost and environmental impact.

In general, the research on the comprehensive evaluation of economic cost and environmental impact within the life cycle in China is still in the stage of independent research, lacking the integration and comprehensive use of the two.

Life cycle economic and environmental impact assessment analysis

Life cycle cost analysis

RealCost, an LCCA software developed by the US federal highway administration, has been recognized and used in a number of states in the United States, becoming a widely recognized LCCA evaluation software. RealCost's LCCA method divides the life cycle cost into two parts, owner cost and user cost [31].

Owner cost

The owner cost is the cost borne by the operator of the pavement. In the range of life cycle, it includes the initial construction cost, maintenance cost, and pavement management cost. The economic costs associated with these processes are attributed to the owner's costs, which can be calculated by the budget method. It must be noted that the calculation range is the life cycle of the road, so it is necessary not only to calculate the economic cost of the whole process of construction acceptance but also to estimate the economic cost of daily maintenance, rehabilitation, and recycling after the road is put into use as well as the economic value in the end of the road life cycle. Here is a simple example of how this economic value is calculated:

Suppose that the life cycle economic cost of two different road schemes needs to be evaluated and the time range of evaluation is 30 years. If one of the roads just reaches its service life in the 30th year, then the second kind of residual value is 0. If one of the highways is reconstructed in year 28 and will remain in use until year 35 to reach its useful life, the size of the second type of remnant is.

$$(35-30)/(35-28) \text{ x the cost of the road rehabilitation} \quad (1)$$

User cost

User cost can be divided into three parts: vehicle operation cost, delay cost, and safety cost. Vehicle operation cost refers to the cost of vehicle operation near the maintenance operation area, and its size is affected by such factors as vehicle type, vehicle age, and the condition of maintenance operation area. Delay cost refers to the time delay cost caused by the maintenance operation area, whose main part is the person in the vehicle. Therefore, its size is not only affected by the nature of the construction operation area but also closely related to the time cost of the person. Safety cost refers to the cost of additional accidents caused by the presence of the maintenance operation area. The detailed calculation models will be mentioned in the following part of "Independent algorithm of LCCA."

The existing evaluation method of the economic cost of life cycle is relatively complete, which mainly takes into account the construction and repair costs of the owner in the life cycle and the economic, time, and safety costs of the user in the construction process and converts them into a unifi d economic indicator through the discount rate.

Life cycle assessment analysis

The LCA method can be divided into three categories, namely, process-based LCA (PLCA), input–output LCA (I-OLCA), and hybrid LCA (hLCA) [32].

Process-based LCA (PLCA) is derived from the study of Coca-Cola bottles by the Midwest Institute of the United States in the late 1960s and is the earliest and most traditional method of life cycle assessment. The latest standard ISO 14040/ISO 14044 issued by the ISO in 2006 established the basic framework of LCA and proposed related requirements and guidelines [33]. It is an analytical method, mainly through investigation and literature review, collecting the input and output lists during the product life cycle.

Unlike PLCA, the input–output life cycle assessment (I-OLCA) is a method of pursuing an overall life cycle analysis. It fi st uses the input and output of the entire department to calculate the energy consumption and emission levels at the

Name	Advantages	Disadvantages
PLCA	• Strong pertinence. • It can compare the environmental impacts of different products according to the specific product type and model, and can adjust the inventory according to the specific conditions of the product, and obtain targeted evaluation data and results [34].	• Errors are inevitable due to the subjectivity of the system boundary and boundary conditions. • Many personal factors may affect the evaluation process as well.
I-OLCA	• The boundary of the input-output of the department is the whole social system, there is no error on the system boundary, which is superior to PLCA.	• It is only possible to obtain the departmental average level and the access to detailed data is difficult. • Not as targeted as PLCA.

Table 1.
Comparison of LCA methods.

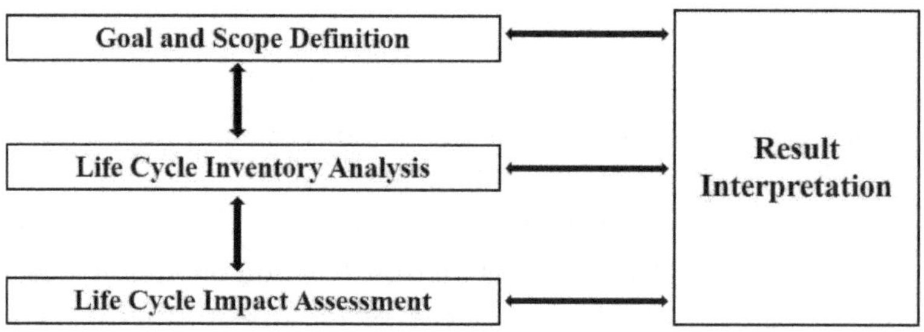

Figure 2.
Procedure of PLCA. These four steps are interrelated and interacting. For example, the problems discovered during the interpretation phase can be returned to the impact assessment, inventory analysis, and even the goal and scope determination steps to be corrected [34].

department level and then evaluates the environmental impact of specific products through the corresponding relationship between the evaluation target and the economic sector. The advantages and disadvantages of each method are shown in **Table 1**.

Hybrid LCA (HLCA) is the combination of the above two methods for evaluation, which makes it possible to eliminate the error caused by the system boundary and enhance the pertinence of the evaluation object, so it is more widely used. However, due to the complexity of the road system, its input and output are both diversifi d and recessive. Therefore, the existing road life cycle assessment is mainly based on PLCA. If not explained in detail, all life cycle assessment methods in this chapter refer to the PLCA.

The life cycle assessment method can be divided into four steps according to the ISO standard: determining the goal and scope, inventory analysis, impact assessment, and result interpretation, as shown in **Figure 2.**

Integrated life cycle economic and environmental impact method

As mentioned above, the calculation process of life cycle cost analysis is to divide the total cost into two categories according to the undertaker, owner cost and user cost, and further subdivide and calculate these two types of costs. The road life cycle inventory analysis process, in order of time and space, calculates the environmental impact of the whole life cycle; however, further discussion of each part will fi d a lot of similarities between the objects evaluated by the two as shown in **Figure 3.**

For example, the "raw material acquisition" and "construction" stage in LCI and the "construction" part in LCCA are evaluated on the pavement materials and construction process. The "maintenance" stage in LCI includes both the "maintenance" and "management" processes in LCCA. It refers to the maintenance work performed by the owner to maintain its structure and function after the pavement is put into use. It also contains the user's cost of evaluation, which is the additional cost and impact of the maintenance of the user. Therefore, there is a great deal of consistency between LCI and LCCA in the process of evaluation, which is also because both of them take pavement as the evaluation object. There are overlaps between the two methods. Many calculations are done by budget method. Both of them are the selection methods of multi-plan comparison, highlighting the differences of multi-plan while downplaying or ignoring the evaluation of the similarities of multi-plan. The biggest difference between the two lies in the different evaluation objectives: LCCA aims at the economic cost, while LCI aims at the environmental impact.

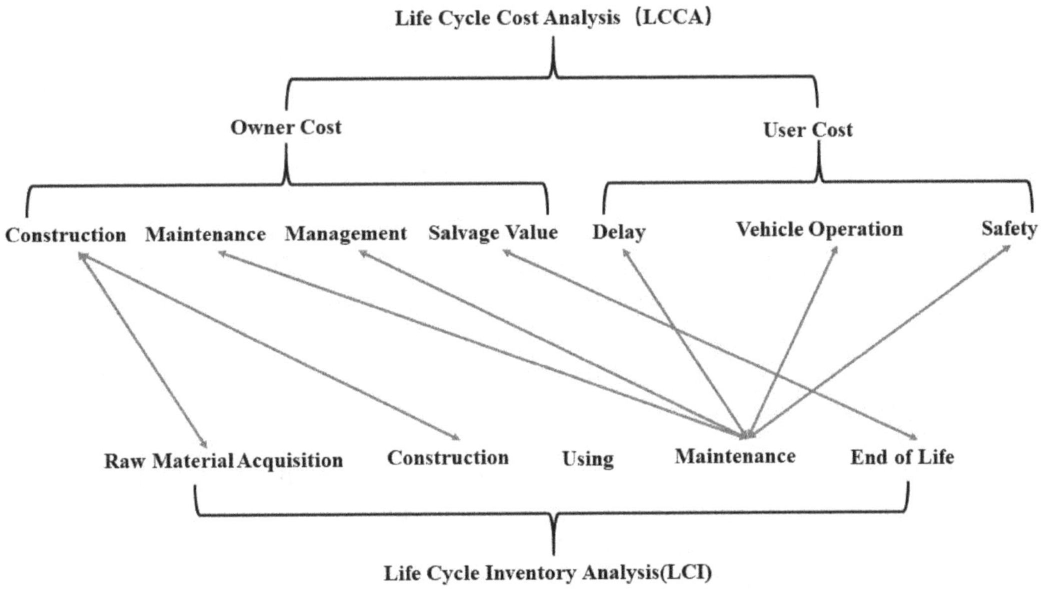

Figure 3.
Differences and similarities between LCA and LCCA.

Since LCCA and LCI have a lot of similarities between the process and the framework, it is possible to take both LCCA and LCI into account. Compared with LCCA's classification method based on undertaking subject, LCI's spatiotemporal sequence is relatively easier to understand and operate, and LCA's framework is also more extensive and logical. Therefore, it is possible to integrate LCCA's evaluation goals into the process of LCI to realize synchronous analysis of economic cost and environmental impact.

Goal and scope

The research goal that should be identifi d includes the cause of the research, the intended use of the research results, the intended users, and publicity; the scope of the study to be determined includes the research object and its functional units, system boundaries, boundary conditions, impact assessment methods and categories, interpretation methods, assumptions, limitations, and other various research elements. The research objectives vary according to the collective situation of the evaluation, while the research scope such as functional units and system boundaries have their commonalities.

Inventory analysis

The inventory analysis step is to make statistics and calculations of the environmental impacts in each stage of the pavement life cycle, including data collection and data calculation.

Due to the complexity and protracted nature of the pavement system, this process is generally divided into several stages. The common practice divides the whole life of the pavement into five stages: raw material acquisition stage, construction stage, using stage, maintenance stage, and end of life, as shown in **Figure 4**.

Raw material acquisition stage

The inventory analysis of the raw material acquisition stage mainly calculates the environmental impact of all pavement material production processes before

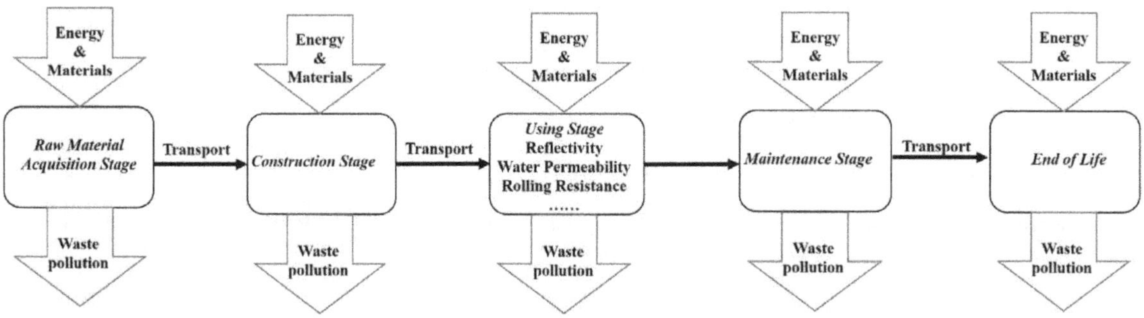

Figure 4.
General pavement life cycle.

construction. Th s process includes not only the environmental impact of the production process of materials such as asphalt, cement, and aggregates but also the transportation and mixing processes of these materials [35]. The environmental impact calculation method at this stage is similar to the budget estimate. The overall environmental impact is calculated via the product of the amount of material and equipment used and the environmental impact per unit amount. The environmental impact per unit can be determined by the product of energy consumption per unit and the environmental impact of energy combustion per unit.

Taking carbon emissions as an example, if there are **n** species of energy and **m**(i) types of material or equipment that consume the **i**th energy, the total carbon emissions can be calculated using Eq. (2) along with Eq. (3). Other kinds of impacts can be calculated in a similar way:

$$\text{Total carbon emissions} = \sum_{i=1}^{n} \text{Energy consumption (i)} \times \text{Carbon emissionsper unit} \tag{2}$$

$$\text{Energy consumption(i)} = \sum_{j=1}^{m(i)} \text{Materials or equipments consumption (j)} \times \text{Energy consumption per unit} \tag{3}$$

The production process of some materials is relatively complicated, and the environmental impact per unit is difficult to obtain quickly. For example, asphalt as a petrochemical product is one of the many products in the petrochemical industry. Environmental impacts need to be further counted and distributed to each product, and its production process varies with region and time, making it difficult to obtain an accurate value [36]. Therefore, the collection of these data is difficult to achieve through individual behavior, and it requires the efforts of governments and organizations.

Countries such as Europe and the United States have been working on this aspect earlier and have obtained a lot of relatively reliable data. In developed countries, work in this area is mostly carried out and supervised by industry associations such as asphalt associations and concrete associations, and they, respectively, investigate and collect environmental impacts of products in their industry [37]. It is worth noting that the data collection and environmental impact assessment methods coordinated by industry associations are not necessarily input–output LCA methods. In fact, they use PLCA more.

Construction stage

Th s stage mainly calculates the environmental impacts of pavement leveling, spreading, and rolling. In addition, the transportation of raw materials from the place of origin to the mixing plant and transportation from the mixing plant to the construction site are all related to this stage and can also be classifi d into this stage. The environmental impact calculation method at this stage is similar to the raw material acquisition stage. The overall environmental impact is calculated via the product of the amount of material and equipment used and the environmental impact per unit amount. The specific energy consumption can be calculated according to the one-shift quota and one-shift consumption of the construction code [38] and can also be analogized according to the actual situation of similar projects.

Using stage

Th s stage mainly calculates the environmental impact caused by the interaction of the pavement surface with vehicles and the environment. It is the most complex phase of the pavement life cycle and the most imperfect stage so far. The pavement system as part of the entire transportation system, its performance, and behavior will have an impact on the environmental burden of vehicles and the environment [39]. For these impacts, many researchers have studied the specific influence modes and relationships from various aspects, among which the research on road rolling resistance and refl ctivity is especially numerous. The following mainly introduce the environmental impact model of the pavement surface from two aspects of rolling resistance and refl ctivity.

Rolling resistance impact model

The rolling resistance of the pavement is the main factor affecting the vehicle consuming during the interaction between people and vehicles. There are now many models for assessing the impacts of rolling resistance on vehicle fuel consumption, which can be divided into four categories depending on whether rolling resistance changes and vehicle speed changes are considered. The more factors are included, the higher the model's simulation of the real situation and the more complex the relative model. Commonly used models include the HDM-4 model issued by the World Bank [40] that considers variable rolling resistance and constant speed and MOVES model for variable rolling resistance and vehicle speed released by the US Environmental Protection Agency [41].

Wang of the University of California Pavement Research Center (UCPRC) proposed a comprehensive rolling resistance environmental impact assessment method in 2012 [39–42]. Based on Wang's research on rolling resistance [36],

$$F_{rolling}=CR_2 \times FCLIM \times (b_{11} \times N_w + CR_1 \times (b_{12} \times M + b_{13} \times v^2)) \tag{4}$$

where $F_{rolling}$ is the rolling resistance (N); CR_1 is tire type parameter; CR_2 is pavement characteristic parameter related to international roughness index (IRI), mean texture depth (MPD), and defl ction value; *FCLIM* is the climatic factor; N_w is the total number of tires; b_{11}, b_{12}, and b_{13} are parameters about tire type and technique; M is the vehicle quality; and v refers to vehicle speed (m/s).

Then use the MOVES model to calculate the relationship between rolling resistance and fuel consumption [35]:

VSP = Rolling resistance þ Air resistance þ Inertial and Gradient resistance

$$\begin{aligned}
&= F_{rolling} \times \frac{v}{M} + F_{Aerodynamic} \times \frac{v}{M} + F_{\text{Inertial and Gradient}} \times \frac{v}{M} \\
&= C_{R}g \times v + \frac{1}{2}\frac{\rho_a C_D A_{front}}{M}(v+v_w)^2 \times v + (a(1+\varepsilon_i) + g \times grade) \times v \\
&= \frac{A}{M} \times v + \frac{B}{M} \times v^2 + \frac{C}{M} \times +(a(1+\varepsilon_i) + g \times grade) \times v
\end{aligned} \tag{5}$$

where VSP is vehicle-specific power that refers to vehicle power per unit mass (W/kg), $F_{Aerodynamic}$ is air resistance (N); $F_{\text{Inertial and Gradient}}$ is inertia or gradient resistance (N); C_R is rolling resistance coeffici t; ρ_a is ambient air density (1.207 kg/m^3, 20°C); v is vehicle speed (m/s); v_w is vehicle upwind speed (m/s); A_{front} is vehicle windward area (m^2); *CD* is air resistance coeffici t; ε_i is the quality factor, its value equal to the equivalent translation quality of rotating components (wheel, gear shaft, etc.) in the transmission system; *grade* is the gradient which is the vertical rise divided by slope length; g is acceleration of gravity (m^2/s); M is vehicle quality (kg); a is vehicle acceleration (m^2/s); A is the rolling resistance coeffici t in the MOVES model; B is the high rolling resistance and rotational loss coeffici t in the MOVES model; and C is the air resistance coeffici t in the MOVES model.

The specific power of a vehicle can be used to measure the power required to operate a vehicle under different conditions, and together with the speed of the vehicle determines the state and fuel consumption of the vehicle engine. The MOVES model simulates the operating state of each vehicle in a certain time range by calculating the specific power and speed of the vehicle running every second, and then sums the time and the number of vehicles according to the state and fuel consumption of different vehicles, fi ally obtain the overall fuel consumption of the vehicle in a certain time and space.

The MOVES model uses a simulation method to calculate the fuel consumption of a large number of vehicles which is relatively accurate and meticulous, but there are also many problems in its local application. First of all, due to the existence of a large number of environmental impact assessment method, this section adopts a simplifi d calculation method for the influence of rolling resistance on fuel consumption, which is easy to operate:

First, according to Wang, the linear relationship between vehicle fuel consumption and IRI is introduced in Eqs. (6) and (7):

Additional fuel consumption of gasoline vehicle = (IRI–Initial IRI) × 0:0313 × Length × Standard fuel consumption of gasoline vehicle × Traffic olume × Length of the road (6)

Additional fuel consumption of diesel vehicle v = (IRI–Initial IRI) × 0:00739 × Length × Standard fuel consumption of diesel vehicle × Traffic olume × Length of the road (7)

Then, according to the IRI decay formula and maintenance formula, the continuous pavement parameters in a certain time can be obtained:

$$\sqrt{IRI} = -0.174 + 9.66 \times 10^{-5} \times \sqrt{CumulativeESAL} + 1.15 \times \sqrt{InitialIRI} \tag{8}$$

$$IRI \text{ change} = -0.6839 + 0.6197 \times \text{Initial } IRI \tag{9}$$

where IRI refers to the international roughness index of the pavement at any time (m/km), cumulative ESAL refers to the cumulative axle load frequency after maintenance, and Initial IRI is initial international roughness index after road maintenance.

Reflectivity impact model

The refl ctivity of the pavement refers to the refl ction ratio of the road surface to solar radiation. The refl ctivity of the pavement affects the surrounding environment in various ways, thereby generating economic cost and environmental impact.

Lawrence Laboratories in the United States released their refl ctivity model in 2017. It takes urban building energy consumption as the evaluation object and evaluates the environmental impact of refl ctivity from a city perspective [43]. Increasing the refl ctivity of the pavement reduces the amount of heat absorbed by the pavement and increases the amount of heat that is refl cted to the surrounding buildings. The former reduces the average temperature of the city and alleviates the urban heat island effect; the latter increases the temperature of nearby buildings, increases cooling costs, and reduces heating costs. In general, the former has a greater utility than the latter, so a highly refl ctive pavement can effectively alleviate the urban heat island effect. There are also many studies that assess the environmental impact of road refl ctivity from a more macro perspective, considering the effect of refl ctivity on radiative forcing. Radiative forcing is a measure of the extent to which a factor affects the earth-atmosphere system's energy ingress and egress energy balance. It is also an index that refl cts the importance of this factor in the underlying climate change mechanism. There are many ways to calculate radiative forcing, and the simplest one can be calculated using Eq. (10) [44]:

$$\Delta mCO_2 = 100 \times C \times A \times \Delta\alpha \tag{10}$$

where Δm_{CO2} is the amount of change in CO_2 emissions; C is a constant of CO_2 emissions, using 255 kg/m^2 as the reference value; A is pavement area; and $\Delta\alpha$ is the variation of pavement refl ctivity. Th s model considers only the effect of refl ctivity changes on CO_2 emissions, so it is the simplifi ation model without considering time and environment.

In addition, there is another method for calculating the radiative forcing considering time variation, as shown in Eq. (11) [45]:

$$+0.01\alpha = \frac{1.087 \times RF \times t}{0.217 \times t - 44.78e^{-t/172.9} - 6.26e^{-t/18.51} - 0.22e^{-t/1.186} + 51.26}[kgCO_2] \tag{11}$$

The left side of this formula indicates a change in the refl ctance per unit area of 0.01, and the right side indicates the CO_2 emissions caused by the change in refl ctance over time t. RF refers to the change in radiative forcing due to changes in surface refl ctance, with a reference value of 1.12–2.14 W/m^2. Because this method is relatively simple, it does not need to consider the localization of multiple parameters. It has been used by many studies and indirectly proves that it has certain reliability.

Impact from other factors

In addition to the above factors, cement and asphalt binders will also undergo changes in properties under environmental influences, which will have an impact on the environment. During the fi ing of cement, limestone releases a large amount of CO_2. With the long-term use of cement pavement, the limestone in the pavement will reabsorb the CO_2 in the air. Th s process gradually reduces the concentration of CO_2 in the air and forms a negative carbon emission value. However, since the speed of absorbing CO_2 is difficult to determine, this process may take several years or maybe decades or centuries [46]. In the long-term use of asphalt pavement, there will be surface runoff on the densely paved road surface, and there will be permeate water on the permeable pavement, which will bring the asphalt precipitate in the asphalt mixture into the water source. However, many studies have shown that it is unlikely that pollutants in the asphalt pavement will reach dangerous concentrations [47].

Maintenance stage

Th s stage mainly calculates the environmental impact of various maintenance strategies during the long-term use of the pavement. The main environmental impacts at this stage are divided into direct and indirect effects. Direct impacts include the environmental impacts of material production and maintenance construction required for maintenance activities, which are similar to the material production and construction phases. Indirect impact refers to traffi delays caused by maintenance activities, which creates an additional environmental burden. The maintenance of the pavement must partially or completely block traffi for a period of time, causing the vehicle to slow down or bypass, which will result in an increase in fuel consumption of the vehicle.

End of life

Th s stage mainly calculates the environmental impact caused by different treatment methods at the end of the life of the pavement. The main disposal methods are classifi d into two categories: burying and recycling [48].

The disposal method of burying is to crush the pavement material and bury it. The environmental impact of this process is divided into three parts, namely, the consumption of crushing, transportation, and burying. There is little literature on the environmental burden of materials after burying, and further research is needed.

Recycling is to break up the pavement material and use it as aggregate to be added to the new pavement material in a certain proportion. In actual engineering, there are various methods for recycling, which can be divided into thermal regeneration and cold regeneration depending on the regeneration temperature and can also be divided into on-site regeneration and in-plant regeneration according to the regeneration site. Since the recycled material comes from the old pavement system and is used in the new pavement system, how the environmental benefits brought by the circulation are distributed between the two systems is a problem still being studied and discussed. The existing distribution methods include cutoff, loss of quality, closed loop, equalization (50/50), and substitution [44]. But there is still no way to get consistent recognition. Due to the lack of data, the equalization method is the most commonly used method. Although it ignores the quantity and importance of recycled materials, it has the best maneuverability in practice [49].

Independent algorithm of LCCA

Labor costs and direct monetary inputs

There will be a large amount of labor input in the process of road construction, maintenance, and recycling. At the same time, some direct monetary input as indirect fee is inevitable. The economic costs of these inputs are relatively easy to calculate, but their environmental costs are difficult to measure directly, so they are calculated as independent economic costs, regardless of their synchronous environmental costs. In the actual calculation, these costs will be directly incorporated into the total economic costs of the corresponding stage.

User cost in the maintenance stage

As mentioned in the analysis of LCCA, user cost can be divided into three parts: vehicle operation cost, delay cost, and safety cost [2]. Vehicle operation cost refers to the cost of vehicle operation near the maintenance operation area, and its value is affected by vehicle type, vehicle age, condition of maintenance operation area, and other factors. The calculation formula of vehicle delay cost in the maintenance area is as Eq. (12), where the vehicle operation cost (/km·vehicle) is a value that changes with areas and time:

$$\text{Vehicle delay cost} = \text{Length of operation area} \times \text{AADT} \times \text{Duration of operation} \times \text{Vehicle operation cost} \quad (12)$$

Delay cost refers to the time delay cost caused by the maintenance operation area; it is not only affected by the construction operation area but also closely related to the time cost of people. The calculation of delay cost is shown from Eqs. (13) to (16). The time value is determined by the average income level and working hours:

$$\text{Deceleration delay time} = (\text{Length of operation area/Speed of operation area}) - (\text{Length of operation area=Upstream driving speed}) \quad (13)$$

$$\text{Queue time} = \text{Queue length/Queue speed} \quad (14)$$

$$\text{Total delay time} = \text{Deceleration delay time} + \text{Queuing time} \quad (15)$$

$$\text{Delay cost} = \text{Total delay time} \times \text{AADT} \times \text{Working time} \times \text{Time value} \quad (16)$$

Safety cost refers to the cost caused by additional accidents due to the existence of maintenance operation area. The specific alculation method is shown in Eqs. (17) to (20). The parameters refer to the values in **Table 2:**

<table>
<tr><th></th><th>Percentage difference of accident rate in operation area (%)</th><th>Number of accidents per million kilometers</th><th>Unit accident cost ($)</th></tr>
<tr><td>Death</td><td rowspan="2">45%</td><td>0.9</td><td>2275229</td></tr>
<tr><td>Injury</td><td>57.2</td><td>15151</td></tr>
</table>

Table 2.
The value of safety cost parameter.

$$\text{Vehicle mileage} = \text{Number of daily trips} \times \text{Days} \times \text{Mileage} \tag{17}$$

$$\text{Number of accidents} = \text{Number of accidents per km} \times \text{Vehicle mileage} \tag{18}$$

$$\begin{aligned}&\text{Number of accidents caused by operation area} = \text{Number of accidents} \\ &\times \text{Percentage difference of accident rate in operation area}\end{aligned} \tag{19}$$

$$\text{Safety cost} = \text{Number of accidents caused by operation area} \times \text{unit accident cost} \tag{20}$$

Economic cost discount rate

When discussing the costs generated in the future, the expected costs and benefits in the future need to be converted into present value, which is called net present value (NPV). Due to the uncertainty in the future, the conversion of future benefits and costs into present value will show the trend of depreciation, and the degree of depreciation depends on the economic and social environment, represented by the discount rate. Since the life cycle cost analysis of the road only considers the cost expenditure, the calculation formula of its NPV is as follows [2, 3]:

$$\text{NPV cost} = \text{Initial construction cost} \sum_{k=1}^{N} \text{k th Future expected cost}$$

$$\left[\frac{1}{(1 + \text{Discount Rate})^{\text{years expected}}}\right] \tag{21}$$

Impact assessment

So as to use the results of the inventory analysis for decision-making, the results of the inventory analysis must be collated and compared to illustrate the equivalent value and importance of each specific environmental impact category. Th s evaluation process can be divided into four steps: classifi ation, distribution, characterization, and quantifi ation.

Classification

Classifi ation is to put the result of the inventory analysis into different environmental impact categories. Th s phase requires the selection of appropriate classifi ation methods and models to distinguish different impact categories. The current classifi ation is to divide environmental impact into three categories: resource consumption, natural environment, and human health, with subgroups of abiotic resource use, acidifi ation, climate change, eco-toxicity, eutrophication, human toxicity, land use, particulate matter formation, photochemical ozone formation, stratospheric ozone depletion, water use, etc. [50].

Assignment

The assignment assigns the results of the inventory analysis to each category to determine the impact category for each of the output substances. Many pollutants can be classifi d into the same category. For example, nitrogen oxides (NO_x) and sulfur oxides (SO_x) can be classifi d into acidifi ation, and carbon dioxide (CO_2) and methane (CH_4) can be classifi d as climate change, as shown in **Figure 5.**

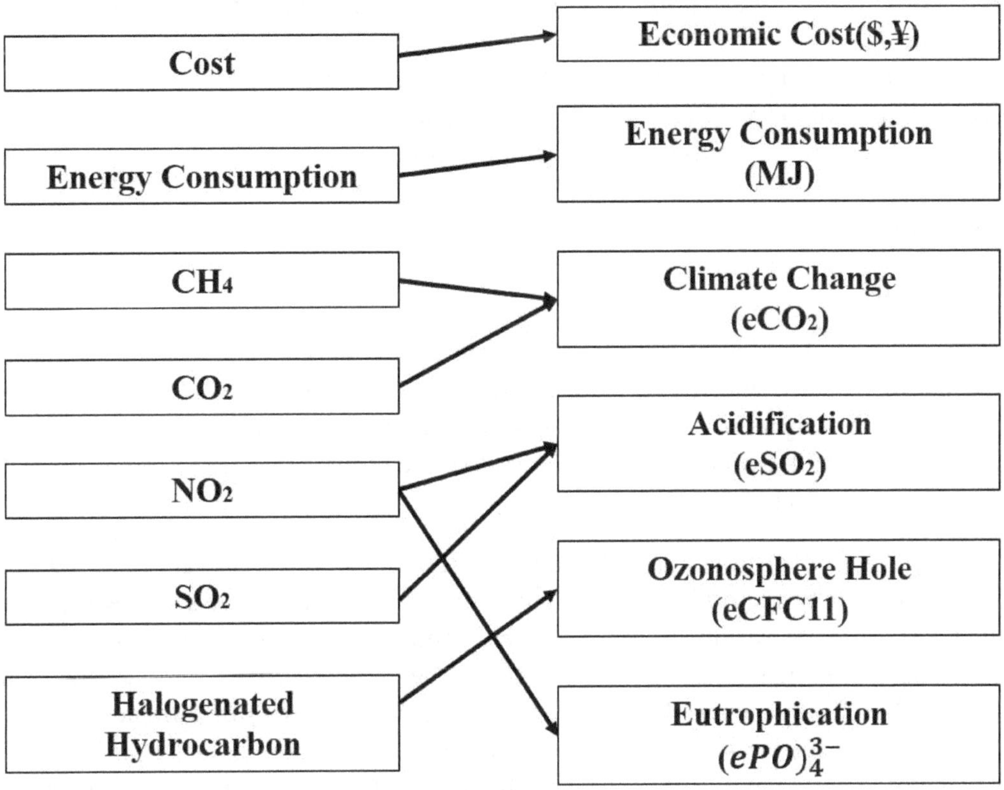

Figure 5.
Classification, assignment, and characterization.

Characterization

The process of characterization is to attribute different pollutants in the same category to the same indicator. For example, climate change includes not only carbon dioxide (CO_2) and methane (CH_4) but also hydrofluorocarbons (HFCs), perfluorocarbons (PFCs), sulfur hexafluoride (SF_6), etc. Th ough the results of natural science research, the global warming capacity of various greenhouse gases over a certain period of time (generally 100 years) is compared with carbon dioxide, thereby converting them into a certain amount of equivalent carbon dioxide, and summing up, the equivalent carbon dioxide emission (CO_2e) is used to evaluate the global warming potential [51]. Th s convert is to multiply greenhouse gas emissions by a parameter that characterizes global warming to get equivalent CO_2 emissions, and this parameter is called impact factors (IF). For other environmental impact categories, of course, there are corresponding evaluation indicators and impact factors.

Some commonly used characteristic units are given in **Figure 5**, such as evaluating acidifi ation potential by equivalent sulfur dioxide emissions and ozone hole potential by equivalent Freon emissions. It is noteworthy that the characteristic unit is not exclusive, some models evaluate the acidifi ation threat by equivalent nitrogen dioxide emissions, and there are corresponding pollutant impact factors which are different from those used in equivalent sulfur dioxide emissions. Th s is also unmistakable and feasible, and this is the biggest difference between LCA impact assessment models.

Quantification

The process of quantifi ation is the process of data processing of the equivalent indicators of each category. There are two common methods for this process: normalization methods and standardized methods. These two methods are essentially a linear transformation that transforms the data into more easily understood valuescand improves the expressiveness of the evaluation results. The difference between the two is that normalization will classify the evaluation results into the interval [0, 1] and the standardized results are related to the overall distribution of the data. Th s stage is an optional stage in the impact assessment, and the results of the evaluation vary depending on the evaluator and the evaluation method.

Result interpretation

Data uncertainty analysis

When various uncontrollable external factors change, the evaluation plan and conclusion may be affected; this evaluation method is called uncertainty analysis, which is a commonly used method in decision analysis. Th ough this analysis, the impact of uncertainty factors on the evaluation results can be clarifi d and minimized, and the resistance of the evaluation conclusions to certain unforeseen risks can be predicted, thereby verifying the reliability and stability of the scheme.

Knowledge, experience, information, and judgment of future decision-making are required in uncertainty analysis. The commonly used methods are: (1) The profit and loss value of the scheme, that is, to calculate the different benefits caused by various factors, and the scheme with the largest return is the optimal scheme. (2) The regret value of the calculation scheme. Calculate the difference between the return value and the maximum return value of the scheme adopted due to the misjudgment of uncertain factors, and the scheme with the smallest regret value is the best scheme. (3) The expected value. By using probability to calculate the standard value of the scheme comparison, the scheme with the best expected value is the best scheme. (4) Consider the criteria of decision-making without deviating from the rules [52]. To sum it up, uncertainty analysis can be divided into breakeven analysis, sensitivity analysis, probability analysis, and criteria analysis.

Conclusions

Th s chapter proposes a comprehensive evaluation idea of pavement life cycle economic cost and environmental impact based on the life cycle assessment framework, which is essentially equivalent to the environmental impact assessment method considering economic cost. The advantage of this method is that it considers both the economic cost and the environmental impact of the road and puts them in a unifi d framework for discussion and comparison. The results of com- parison can be given more quickly and clearly in multiple schemes than in the selection, which is helpful for decision-makers to make choices.

Although LCCA and LCA have a large number of overlapped parts, some parts are independent of each other. For example, for the labor input of a certain project, economic inputs such as compensation and insurance must be considered, and it is difficult to quantify the environmental impact of labor input. It would therefore be inappropriate to consider only its economic costs and ignore its environmental impact. Therefore, it is suggested that a more comprehensive LCA system should include and is not limited to:

- Environmental aspects: factors such as global warming, human toxicity, resource depletion, ozone depletion, and eco-toxicity
- Economic aspects: factors such as owner costs and user costs
- Social aspects: factors such as worker income, accident rate, worker social welfare, and social disparity (industry, income, etc.)

Author details

Jiawen Liu[1], Hui Li[1,2*], Yu Wang[1] and Nailing Ge[1]

The Key Laboratory of Road and Traffi Engineering, Ministry of Education, College of Transportation Engineering, Tongji University, Shanghai, China

Department of Civil and Environmental Engineering, University of California Pavement Research Center, University of California, Davis, CA, USA

*Address all correspondence to: hili@ucdavis.edu

References

[1] Zhu Q. Pavement Project Life-Cycle Analysis Model and Application. Guangzhou: South China University of Technology; 2010

[2] Walls III J, Smith MR. Life-cycle cost analysis in pavement design. In: Monte Carlo Method. 1999

[3] Delwar M, Papagiannakis AT. Relative importance of user and agency costs in pavement LCCA. In: International Conference on Managing Pavements. 2001

[4] Chan A, Keoleian G, Gabler E. Evaluation of life-cycle cost analysis practices used by the Michigan Department of Transportation. Journal of Transportation Engineering. 2008; 134(6):236-245

[5] Zhu Q et al. Application of real cost for life cycle cost analysis of American Pavement Engineering. Journal of China & Foreign Highway. 2010;30(2):94-98

[6] Landsberg DR, Steward R. Improving Energy Effici cy in Buildings. Vol. 56-61. New York: State University of New York Press; 1980. pp. 290-321

[7] Häkkinen T, Mäke lä K. Environmental impact of concrete and asphalt pavements. In: Environmental Adaption of Concrete. Research Notes 1752. Technical Research Center of Finland; 1996

[8] Horvath A, Hendrickson C. Comparison of environmental implications of asphalt and steel- reinforced concrete pavements. Transportation Research Record. 1998; 1626:105-113

[9] Roudebush WH. Environmental value engineering assessment of concrete and asphalt pavement. In: PCAR&D Serial No. 2088a. Portland Cement Association; 1999

[10] Berthiaume R, Bouchard C. Exergy analysis of the environmental impact of paving material manufacture. Transactions of the Canadian Society for Mechanical Engineering. 1999;23(1B): 187-196

[11] Mroueh UM, Eskola P, Laine Y, et al. Life cycle assessment of road construction. In: Finnra Reports 17/ 2000. Finnish National Road Administration; 2000

[12] Stripple H. Life Cycle Assessment of Road: A Pilot Study for Inventory Analysis. IVLB1210E. Swedish National Road Administration; 2001

[13] Nisbet MA, Marceau ML, VanGeem MG, et al. Environmental life cycle inventory of portland cement concrete and asphalt concrete pavements. In: PCAR&D Serial No. 2489. Portland Cement Association; 2001

[14] Park K, Hwang Y, Seo S, et al. Quantitative assessment of environmental impacts on life cycle of highways. Journal of Construction Engineering and Management. 2003; 129(1):25-31

[15] Zapata P, Gambatese JA. Energy consumption of asphalt and reinforced concrete pavement materials and construction. Journal of Infrastructure Systems. 2005;11(1):9-20

[16] Fang F. Analysis of economy and energy consumption of cement concrete pavement and asphalt concrete pavement. East China Highway. 1984; 01:3-15

[17] Gu Q. Improve pavement smoothness save energy and reduce transport costs. Journal of Highway and Transportation Research and Development. 1986;(01):19-22

[18] Ye G. Optimization design of fatigue and energy consumption of cement-stabilized Macadam base. Journal of South China Construction University (West Campus). 1994;(02): 23-29

[19] Zhang Z. Analysis of ergonomics and energy consumption quota of recycled asphalt pavement. Southwest Highway. 1995;01:23-24

[20] Han C. Economic comparison between cement pavement and asphalt pavement of expressway. Heilongjiang Jiaotong Keji. 2007;(07):131-132

[21] Yi S, Gao Y, et al. Energy consumption calculation and environmental assessment of warm mix and half-warm mix asphalt mixture. Petroleum Asphalt. 2009;(05):74-77

[22] Shang C, Zhang Z, Li X. Research on energy consumption and emission of life cycle of expressway. Journal of Highway and Transportation Research and Development. 2010; 27(8):149-154

[23] Tang Z, Song Y. Application of low energy consumption warm mix asphalt concrete in pavement engineering in Alpine region. Subgrade Engineering. 2011;(05):168-171

[24] Pan M. The Methodology Research and Application on Energy Consumption and Carbon Emissions of Highway based on the Life Cycle Assessment. Guangzhou: South China University of Technology; 2011. pp. 78-79

[25] Shi F, Yu Q, et al. Study on energy consumption of reclaimed asphalt pavement materials by comparison. Road Machinery & Construction Mechanization. 2011;11(45):82-85

[26] Ma F, Qin J, et al. Application of life cycle assessment (LCA) in American highway. Journal of China Foreign Highway. 2014;34(05):332-337

[27] Li X. Environmental Impact Assessment of Cement Pavement and Asphalt Pavement Based on LCA. Southeast University; 2015

[28] Zhang H et al. Effect of design parameters of asphalt pavement based on LCA on carbon emission. Journal of Highway and Transportation Research and Development. 2018;35(02):1-7

[29] Shu Q, Zhang X. Study on the way to integrate life cycle assessment and life cycle cost analysis. Tongji University Journal Social Science Section. 2003; 14(4)

[30] Batouli M, Mostafavi A. Service and performance adjusted life cycle assessment: A methodology for dynamic assessment of environmental impacts in infrastructure systems. Sustainable and Resilient Infrastructure. 2017:1-19

[31] Walls J III, Smith MR. Life-Cycle Cost Analysis in Pavement Design–In Search of Better Investment Decisions. US: Federal Highway Administration; 1998

[32] Wang C, Zhang L, Pang M. A review of life cycle assessment methods— Development and application of the evaluation of hybrid LCA. Journal of Natural Resources. 2015;7:1232-1240

[33] Huo L. Review of life cycle assessment (LCA). China Packaging. 2003;1:42-46

[34] International Organization for Standardization (ISO). Environmental Management—Life Cycle Assessment— Principles and Framework. ISO Standard 14040. Geneva, Switzerland: International Organization for Standardization; 2006

[35] Cai R. Research on Quantitative Analyzing System on Energy Consumption and Carbon Emission of Asphalt Mixtures. Xi'an: Chang'an University; 2013

[36] Kendall A, Harvey J, Lee I-S. A critical review of life cycle assessment practice for infrastructure materials. In: US–Japan Workshop on Life Cycle Assessment of Sustainable Infrastructure Materials, Sapporo, Japan. 2009

[37] European Bitumen Association. Life Cycle Inventory: Bitumen; 2011

[38] Zhu H. Typical Asphalt Pavement Construction Energy Consumption and Carbon Emission Quota Calculation Research Report. Jiangsu: Jiangsu Transportation Institute; 2013

[39] Wang T, Harvey J, Jones D. A Framework for Life-Cycle Cost Analyses and Environmental Life-Cycle Assessments for Fully Permeable Pavements. California: University of California Pavement Research Center; 2010

[40] Zaabar I, Chatti K. Calibration of HDM4 models for estimating the effect of pavement roughness on fuel consumption for US conditions. In: Transportation Research Board 2010 Annual Meeting; Washington, DC. 2010

[41] USEPA. MOVES (Motor Vehicle Emissions Simulator). United States Environmental Protection Agency; 2010

[42] Wang T, Lee I-S, Harvey J, Kendall A, Lee EB, Kim C. UCPRC Life Cycle Assessment Methodology and Initial Case Studies on Energy Consumption and GHG Emissions for Pavement Preservation Treatments with Different Rolling Resistance. California: University of California Pavement Research Center; 2012

[43] Levinson R, Harvey J, et al. Life- Cycle Assessment and Co-Benefits of Cool Pavements. California Air Resources Board; 2017. p. 4

[44] Yu B, Lu Q. Life cycle assessment of pavement: Methodology and case study. Transportation Research Part D – Transport and Environment. 2012;17(5): 380-388

[45] Yu B, Lu Q. Estimation of albedo effect in pavement life cycle assessment. Journal of Cleaner Production. 2014;64: 306-309

[46] Lagerblad B. Carbon Dioxide Uptake During Concrete Life Cycle— State of the Art. Stockholm, Sweden: Swedish Cement and Concrete Research Institute, CBI; 2006; Nordic Innovation Centre Project Number 03018

[47] Brandt HCA, DeGroot PC. Aqueous leaching of polycyclic aromatic hydrocarbons from bitumen and asphalt. Water Research. 2001;35: 4200-4207

[48] Rajendran S, Gambatese JA. Solid waste generation in asphalt and reinforced concrete roadway life cycles. Infrastructure Systems. 2007;13:88-96

[49] Huang S. Method of Analyzing Life- Cycle Energy Consumption for Pavement Structures. Changsha: Hunan University; 2013

[50] Hauschild MZ, Huijbregts MAJ. Life Cycle Impact Assessment. New York: Springer; 2015. pp. 5-6

[51] Van Dam TJ, Harvey JT, Muench ST, et al. Towards Sustainable Pavement Systems: A Reference Document. Federal Highway Administration; 2015

[52] Wikipedia. Data Uncertainty Analysis [EB/OL]. 2018. Available from: http://wiki.mbalib.com/wiki/%E4% B8%8D%E7%A1%AE%E5%AE%9A%E6%80%A7%E5%88%86%E6%9E%90

3

Energy Efficiency Management: State of the Art and Improvement Potential Analysis with Regard to Cargo Transport by Air and Rail

Corinna Cermak, Erich Markl and Maximilian Lackner

Abstract

Th s review article summarizes the state of the art in energy effici cy (EE) management in air and rail cargo transportation. After an introduction, explanations and defin tions follow around the topic of energy efficiency. The political framework conditions of the European Union (EU) as well as the associated European Union Emissions Trading System are described. In particular, the drive technologies, CO_2 emissions, and fuel-saving options are reviewed.

Keywords: energy effici cy, cargo, air, rail

Introduction

The effects of climate change are global. For many years now, polar ice caps and glaciers have been melting, sea levels are rising, and storms and fl ods are destroying people's habitats. Therefore, it is particularly important to reduce anthropogenic greenhouse gas (GHG) emissions through energy effici cy (EE) [1].

Energy effici cy [2] is gaining more and more attention in the society. The term energy effici cy means using less energy to guarantee the same benefits of output [3].

The reasons for the increasing interest in energy effici cy are the persistent climate change, depletion of fossil fuels, and rising energy prices. The production as well as the use of energy based on fossil fuels is one of the biggest drivers of climate change. The sustainable use of energy or "green energy" is playing an increasingly important role in various decision-making processes for companies and other organizations [4].

The application and implementation of energy effici cy is often the cheapest way to reduce fuel costs and carbon dioxide (CO_2). Of course, the needs of the present are to be satisfi d, but future generations should not be disadvantaged. The most important sectors in terms of energy effici cy include industry, buildings, and transport. Year after year, countless institutes and corporations are researching new technologies to ensure that energy effici cy can be continuously optimized, alongside decarbonization [4, 5].

The European Union (EU) has adjusted its policy framework based on annual CO_2 emissions. In 2005, the EU introduced an emissions trading system for all member states to reduce CO_2 emissions (for details, see

Figure 2, purchase and sale of emission certifi ates) [6–9]. Th s chapter deals with the current state of the art of freight trains and cargo aircraft

Energy effici cy means using less energy to provide the same level of utility. It is therefore one method to reduce anthropogenic (human) greenhouse gas emissions [3].

EE is a universally applicable concept relevant for consumers and industry alike that can be achieved by a more effici t technology, an improved process, or a change of individual behavior. Energy effici cy can, according to the International Energy Agency's (IEA) World Energy Outlook (IEA WEO), "close the competitiveness gap caused by differences in regional energy prices" [10].

In November 1974, the International Energy Agency, an autonomous agency, was founded. Its main mission is to promote energy security among its 29 member countries.

Energy effici cy is the key to ensure a safe, reliable, affordable, and sustainable energy system for the future. It is the one energy resource that every country possesses in abundance and is the quickest and least costly way of addressing energy security and environmental and economic challenges. While energy effici cy policies are becoming a key part of the global energy market, there remains vast untapped-into potential [11].

Energy effici cy means using the same amount of energy to achieve the same utility level. The term energy effici cy itself is therefore very clear and self-explanatory [12].

"Energy intensity" looks at how much energy was needed to get a certain result. The units of energy intensity used are usually given by the primary energy consumption per inhabitant or by the primary consumption per unit of gross domestic product (GDP). Measuring energy effici cy as an energy intensity is basically possible at the macroeconomic level [4]. As an indicator of energy effici cy, the energy intensity of a country is often used for the assessment. Th s is because at a high level it is a proxy measure of energy needed to provide the used energy service (the energy intensity measures the energy needed to provide units of economic value). Moreover, it is very readily available as an indicator, and it is easier to rate or compare countries. If a country has a low energy intensity, it does not necessarily mean that the energy effici cy of that country is also high. Conversely, it should be noted that lower-intensity trends are not necessarily due to effici cy improvements. Energy effici cy contributes to

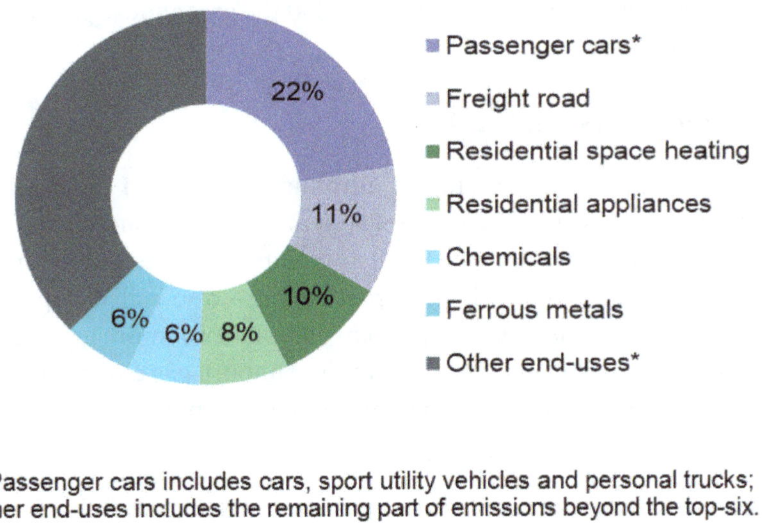

Figure 1.
Top six CO_2-emitting end uses in IEA countries in 2013. Reproduced with permission from [13].

the definition of intensities and trends. However, other factors play a major role, e.g., the structure of the economy, the presence of large energy-consuming industries, the passenger car density, and the specific housing sector. Globally, passenger cars, together with road haulage vehicles, account for about one-third of energy-related CO_2 emissions and consume approximately as much energy as the entire housing sector [13]. Canada, the USA, Australia, and New Zealand are among the regions in which transport is currently the highest consumption sector. The manufacturing sector has the largest share of energy in Japan and Korea. In particular, the subsectors' ferrous metals and chemicals have a high energy demand. The residential sector is the highest in most European countries. This stems very much from the consumption for space heating and heaters. For the IEA countries, emissions for residential heating were higher for household appliances as well as for each subsector of the manufacturing sector. Especially in countries such as the UK and Germany, the heating of the rooms was the largest emitting end use (see **Figure 1** for 2013 data) [13].

Political framework operations in the EU

The European Union has adopted a wide range of legal instruments to improve energy effici cy. For EU directives, it is up to the member states how they transpose the directives into national law. In the offi al journal of the European Union, directive 2009/28/EG promoting the use of energy from renewable sources and amending and subsequently repealing directives 2001/77/EG and 2003/30/EG are binding and non-binding guidelines for an effici t energy use in the EU. All member states have to give their best to implement those improvement measures [6].

The main targets of the directive 2009/28/EG are the so-called 20-20-20 targets. Th s means that, as target, renewable energies account for 20% of the fi al energy consumption across the EU, the greenhouse gas emissions have to be reduced by 20%, and the energy effici cy should be increased by 20% in the EU referring to 1990 as base year. Furthermore, the directive 2009/28/EG stipulates that biofuels must account for 10% of total fuel consumption by 2020 [7].

There are also some long-term objectives of the EU. The proportion of renewable energies in energy consumption and the energy effici cy should increase to at least 27% by 2030. In 2020, the goals should be reviewed to see whether an increase of 30% is possible. In 2030 the greenhouse gas emissions must be reduced by almost 40% (base year 1990) [8].

European emissions trading system (EU ETS)

The European Union Emissions Trading System was introduced in 2005 for the regulation of greenhouse gas emissions in energy-intensive industries. The relevant system operators concerned must have a certifi ate for each tonne of CO_2 emitted and submit it to the regulatory authority. The system enables trading in carbon certifi ates under a defi ed overall framework for all participants—a so-called "cap-and-trade" system. These allowances are freely tradable, which means that participants who emit more greenhouse gases can purchase emission certifi ates from other participants with lower greenhouse gas emission needs. In order to reduce overall emissions, the proportion of allowances is reduced by 1.74% per year by 2020 [9]. **Figure 2** illustrates the EU ETS scheme.

The principle of the "cap-and-trade" system is easy to recognize in **Figure 2**. In this example, issuer A generates more emissions than issuer B, thus exceeding the authorized amount of CO_2. In order to solve the problem, issuer A can purchase the unused certifi ates from issuer B (which has very low CO_2 emissions in this example).

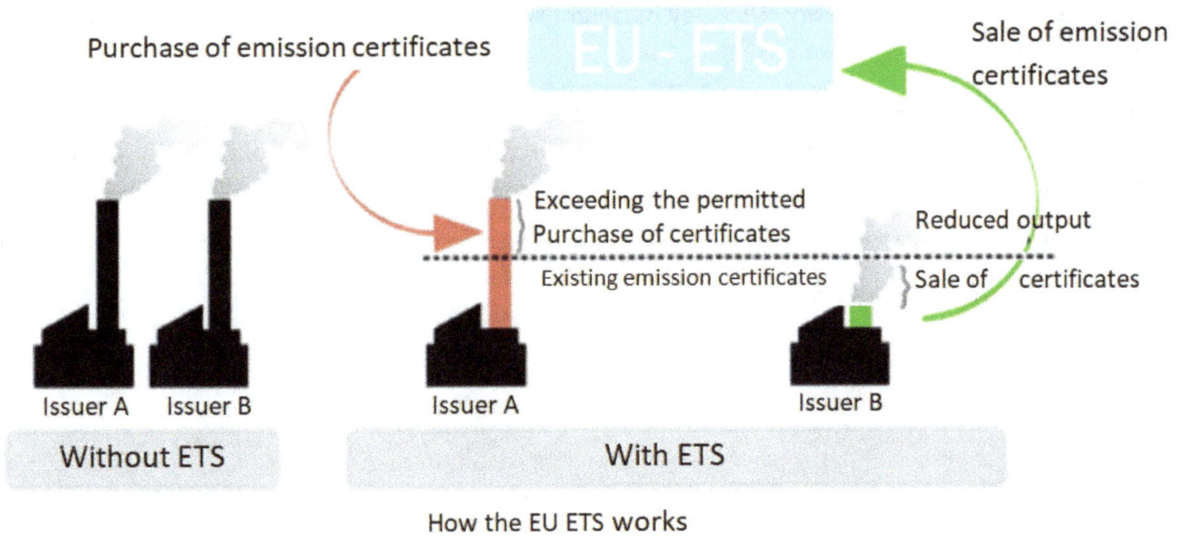

Figure 2.
Purchase and sale of emission certificates. Reproduced with permission from [9].

Energy analysis

In the transport sector, mainly fuels based on crude oil are used, so there is a high dependence on fossil fuels. While diesel fuel showed a steady growth in Europe for several decades, gasoline fuel is now becoming more important again, according to the European Automobile Manufacturers Association (ACEA) [14]. Recently, dual-fuel gasoline/diesel engines were presented [15]. Electric vehicles are strongly on the rise and are expected to replace cars with internal combustion engine in future [16, 17].

The movements of goods and people are very different transport processes and are also recorded differently. It is customary to specify the transport performance in the freight sector in tonne kilometers (tkm) or in person kilometers (pkm). The transport service refers to the distance covered in the respective territory.

The steady growth of online commerce has and will continue to have a major impact on the transport sector in future [18].

Large online mail order companies promote the digitization of trade and thus the amount of goods that have to be transported to the end users. In Germany alone, freight traffi has risen by about 120% since 1990 [19]. Thus, the traffi volume takes up about 26% of the traffic elated energy consumption [20]. The strongest increase is recorded in road and rail transport. For the transport sector, energy intensity indicators are collected on the basis of energy consumption and the transport performance provided.

These energy intensity indicators are used as a statistical measure of energy effici cy. In order to support the sustainable development of economic and livelihood models, indicators in the transport sector should include resource effici cy. Due to the large increase in freight traffi in recent years, the objective within freight transport should be to decouple growth in terms of increased transport performance and CO2 emissions. Energy consumption in freight transport related to rail and aviation is around 15% in aviation and 2% in rail transport, respectively [4].

Overall, it can be summarized that the energy consumption has increased less than the transport performance.

Thus, it can be concluded that energy effici cy in freight transport has increased, but growth is neutralizing the effici cy gains achieved [4].

As it can be seen in **Figure 3**, aviation has the highest specific energy consumption. Low oil prices encourage the (continued) use of older, less energy-effici t cargo aircraft, making them economically viable again or extending their useful

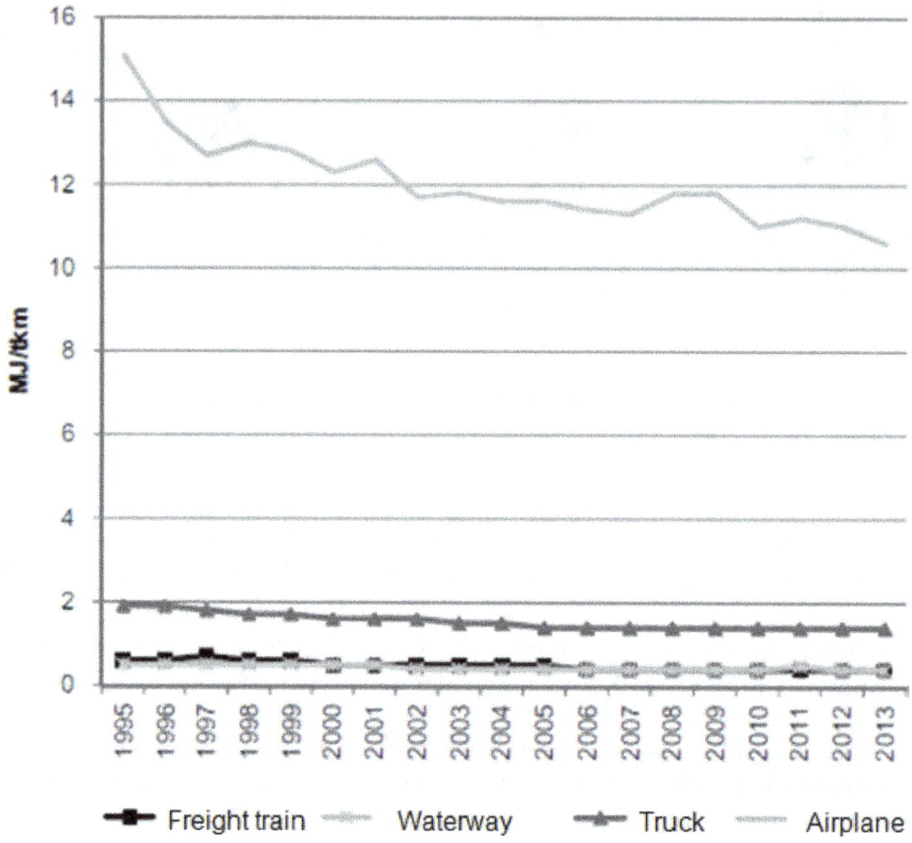

Figure 3.
Development of specific energy consumption in freight traffic by mode of transport from 1995 to 2013 (modified from [21], p. 246).

lifetime. Th s could have an impact on indicator development. In the second place is the use of trucks, which are logistically almost inevitable due to the infrastructure in most countries, particularly for "last mile" delivery.

CO_2 emissions

Atmospheric CO_2, which stands at ~400 ppm, is both harmful and vital. Without the signifi ant greenhouse gas CO_2 and other natural greenhouse gases, the Earth's average surface temperature would be sub-zero, and life would not be possible. Th s natural balance is disturbed by excessive and increasing anthropogenic CO_2 emissions. These disturbances lead to global warming, which has developed since the Industrial Revolution (~1750) and increased signifi antly in recent years [3, 22].

The steadily growing globalization is making a signifi ant contribution to the greenhouse gas effect. Carbon dioxide is also produced in combustion processes of engines and other equipment, with most of the artificial and harmful CO_2 emissions caused by the industrialized countries. In addition to carbon dioxide, e.g., methane (CH_4) and nitrous oxide (N_2O) contribute to the environmental impact. These gases have a higher global warming potential (GWP) than CO_2 [23]. All greenhouse gases can be converted into CO_{2eq} (CO_2 equivalents) [3].

Natural and anthropogenic greenhouse effect

Natural greenhouse gases act as a kind of filter that lets the (short wavelength) sun's rays through and captures the heat radiation (long wavelength) from the Earth's surface, thereby enabling regulated global warming. The functioning of a greenhouse is also used, for example, in plant breeding. Colloquially, the term greenhouse gas effect is used for global warming [22]. The presence of gases with GWP results in an increase in temperature and thus the greenhouse gas effect. Many scientific predictions suggest that the Earth will continue to warm without a reduction in CO_2 emissions [23], e.g., by 2°C when the atmospheric CO_2 concentration can be kept at or below 650 ppm, and signifi antly higher if it cannot [10]. In August 2018, the IPCC has released a special report, pledging to limit global warming to 1.5°C [24]. "Limiting global warming to 1.5°C compared with 2°C would reduce challenging impacts on ecosystems, human health, and well-being," said Priyadarshi Shukla, Chair of the Global Centre for Environment and Energy, Ahmedabad University, India, and the co-author of the report.

Under the 2015 Paris Agreement [25], countries have agreed to keep global temperatures increase below 2°C, which might not be enough to avoid exceeding dangerous tipping points.

The Kyoto Protocol

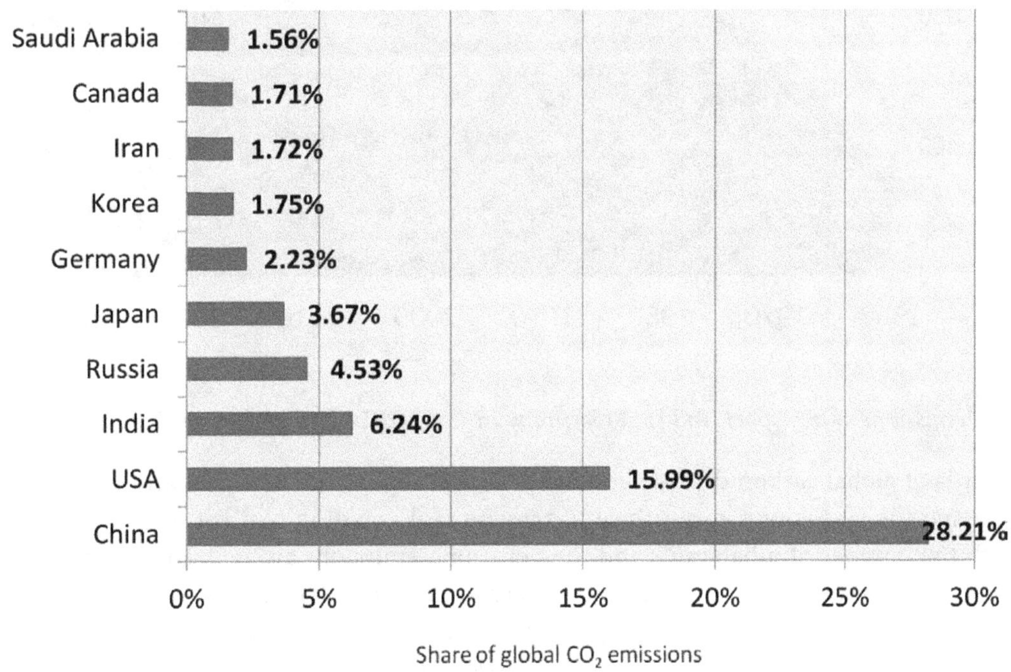

Figure 4.
The 10 countries with the largest share of CO2 emissions worldwide in 2016. Modified from [27].

The Kyoto Protocol [26] is an agreement made in 1997. It regulates the United Nations Framework Convention on Climate Change (UNFCCC) and focuses on climate protection. In doing so, an international and joint agreement was achieved, which should gradually reduce CO_2 emissions worldwide. The Kyoto Protocol is designed to slow down the progressive effects of greenhouse gases and, if possible, to halt artific al global warming [23].

The statistics in **Figure 4** shows the 10 largest CO_2-producing countries by share of global CO_2 emissions in 2016. With a very high proportion of over 28% of global carbon dioxide emissions, China was the world's largest CO_2 emitter in 2016. Second only in the list of climate sinners come the USA with almost 16%. In addition to carbon dioxide, the Kyoto Protocol includes another five greenhouse gases: methane (CH_4), nitrous oxide (N_2O), "F-gases" hydrofluorocarbons (HFCs), perfluorocarbons (PFC), and sulfur hexafluoride (SF_6). Nitrogen trifluoride (NF_3) must be additionally included since 2015. In 2015, Germany was able to reduce almost all greenhouse gases compared to the year 2000 (with the exception of hydrofluorocarbons) and thus already meets the requirements of the Kyoto Protocol [27].

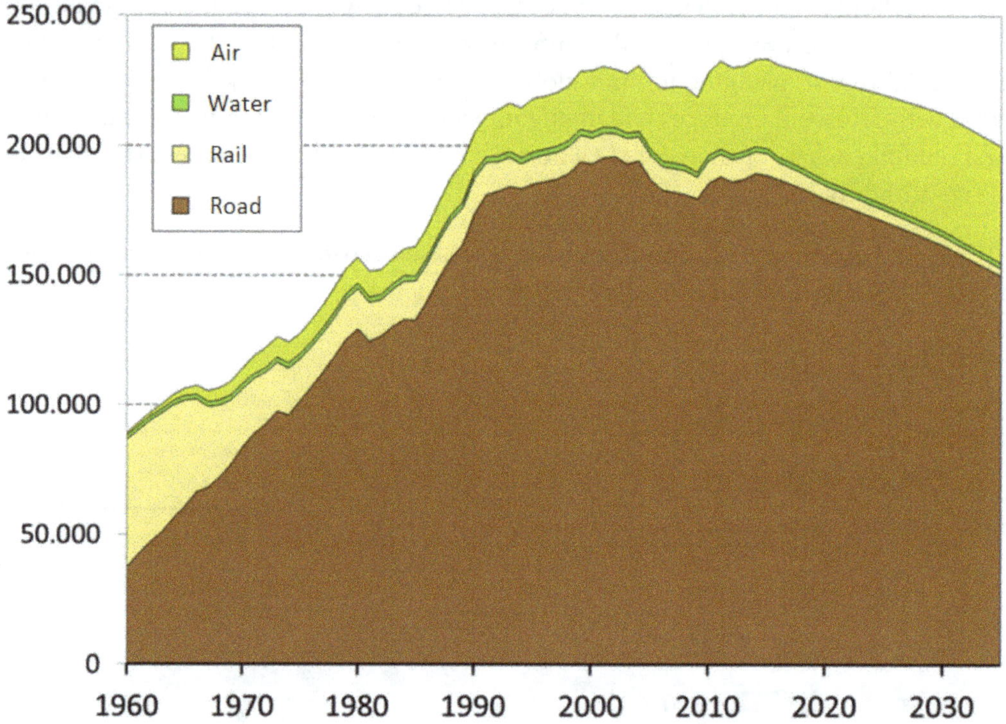

Figure 5.
Carbon dioxide emissions of transport (in kt) in Germany in 1960–2014 and in the trend scenario until 2035 [30].

Nearly two-thirds of global carbon dioxide emissions are caused by 10 countries, around a quarter or one-sixth by China and the USA alone, respectively. Germany ranks sixth in that list. For the total emissions of a country, both the number of inhabitants and the per capita emissions are relevant. In short, an American causes almost twice as much carbon dioxide as a German and about 10 times as much as an Indian citizen. Worldwide, the carbon dioxide emissions and the carbon dioxide concentration in the atmosphere are steadily increasing. Energy-related carbon dioxide emissions have more than doubled worldwide since 1970 and increased more than 15 times since the beginning of the last century. Since 2013, however, emissions have grown signifi antly more slowly. Th s gives a little hope that a stabilization and then a trend reversal can be made possible [28].

The Transport Emission Model (TREMOD) [29] is a commonly used method of determining CO_2 emissions.

The TREMOD takes into account direct emissions from vehicles, evaporative emissions, and total emissions with the energetic upstream chain [4] (compare **Figure 5**).

In the transport sector, not only the specific energy consumption but also the development of CO_2 emissions is an important indicator of energy effici cy. Concerning CO_2 emissions, very different information is given in the transport sector, mainly because of the increased use of specific energy sources and signifi ant differences in the CO_2 balance among the individual modes of transport [4].

Rail freight

According to a study by SCI traffi "Rail Transport Markets-Global Market Trends 2016–2025," global rail freight traffi fell by 4.4% in 2015 [31]. Thus, the rail freight traffi has recorded negative growth for the fi st time since many years. The reason for this is the slowdown in the transport of coal and steel due to weaker production activity and also the decline in international trade growth. In the face of growing protectionism, transport demand is expected to be negatively impacted worldwide. Coal transport in North America has declined signifi antly as a result of the change in the energy mix [31].

In Asia, above all, the Chinese transport market, which is shrinking in rail freight transport [31], dominates. The rising tariffs in rail freight traffi are the cause of the reduction. The strong expansion of road infrastructure with highways, which has made road transport more attractive, also led to a decline in rail traffic However, there is a desire in the Chinese area for a sustainable improvement in air quality, which can also be achieved by shifting the modal split in favor of greener rail. However, it is still unclear whether and to what extent the corresponding political decisions will take place and whether they will be able to compensate for the declining coal transport by rail [31] (compare **Figure 6**).

The basis for rail freight transport is freight wagons in regular or special design. These are coupled to wagon groups (half trains or block trains). Newer trains have a total length of up to 700 meters. Each wagon has a payload of 25–62 tonnes. The fi st railways were driven with the power of steam engines. However, this technique is no longer used today due to the low effici cy. At present, mainly internal combustion engines and electric motors are used for the drive. Here the drive is in locomotives or in the railcar. While a railcar carries the goods to be transported, a locomotive pulls wagons with the goods to be transported [32].

Freight trains are divided into different types of trains:

Block trains: Carriage of goods or a shipment from one customer to another one; between the departure station and the destination station, the whole train remains unchanged. The wagons of the same design are put together. Th s favors standardized loading processes for consignors and consignees as well as equipment that is specially tailored to the type of wagon [33].

Mixed freight wagonload/single wagonload freight trains: Individual wagons are used to build trains for different customers, which must be dismantled and reassembled in shunting yards [33].

Mixed block trains: Mixed block trains stand for trains that consist of several blocklike segments that carry different goods and/or have different routes. Th s is intended to combine the flex bility of wagonload traffi with the effici cy of block train traffic [33

In internal combustion engines (as in everyday traffic) typically commercial diesel fuel is used. By the invention of the electric motor (1837) and generators

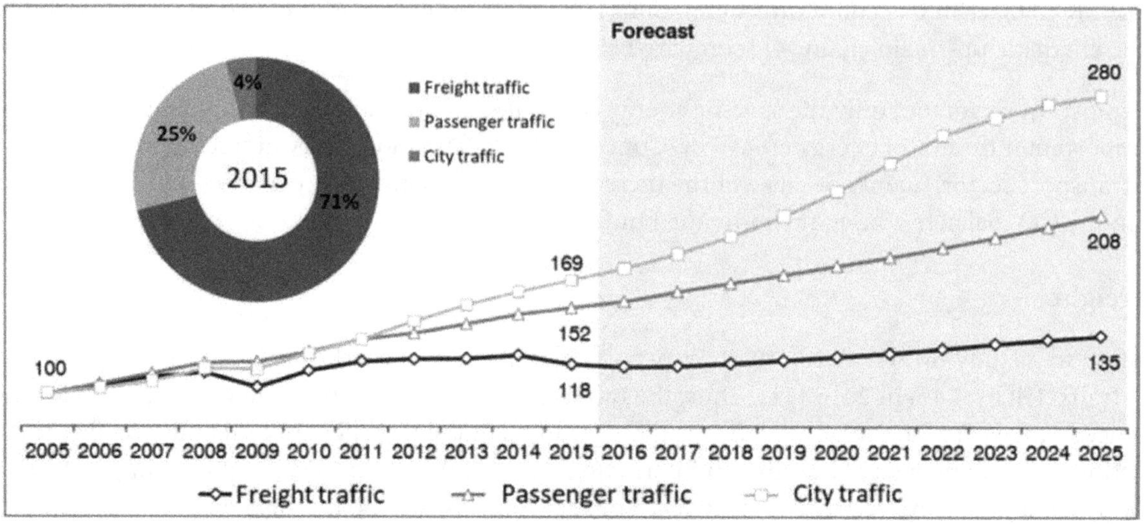

Figure 6.
Worldwide rail services; index 100 (average) = 2005, pkm (person kilometers), tkm (tonne kilometers). Reproduced with permission from [31].

and transformers (1866), it was possible to develop the fi st electrically operated railways. The energy required was initially supplied directly via batteries or via the tracks. AC motors or DC motors are used [32] (AC = alternating current, DC = direct current).

In order to increase energy effici cy, unnecessary transport routes must be avoided. Traffi avoidance can be achieved by means of dynamic route planning by telematics systems. It uses information technologies and communication technologies as well as up-to-date traffi information for tour planning and tour control. Data streams from location, navigation, data and voice communication, and vehicle data are transmitted, collected, processed, and sent back in real time to the vehicles for their control. Th s relieves traffi infrastructure and the environment [32].

Apart from that, measures to optimize capacity utilization are important, since the means of transport are mostly only moderately utilized and empty trips (trips without goods to be transported) are not uncommon. Th s includes, for example, bundling. In "time bundling," future requirements of a target area are combined to transport blocks, while in "spatial bundling," orders from neighboring customers, e.g., by transshipment point or cross-docking, are pulled together [32].

Existing and emerging fuel-efficient locomotives

At the 2012 American Society of Mechanical Engineers (ASME) Joint Rail Conference, technology options for use in new rail systems, retrofits, or systemwide energy effici cy gains were assessed as being proven and highly promising. Among other things, it was recognized that the most important factor in improving energy effici cy is the modernization of the traction and propulsion system. The modernization of the heat, ventilation, and air conditioning (HVAC) system control for railcars allows further system-wide energy efficiency gains [34].

The focus here is on integration:

- Variable fans or dampers.

- Frequency converters for refrigerant compressors, which save energy for heating and ventilation (saving potential can reach up to 70%).
- Permanent magnet motors, which increase the effici cy. They also reduce the size and weight of compressors/pumps.

Improved rail and facility effici cy lighting system options (including stations, depots, and rails, as well as multimodal terminals) include daytime lighting, automatic ambient light sensors, and motion detectors. Furthermore, incandescent lamps and fluorescent lamps are being replaced by long-lasting, low-power, lightemitting diodes (LEDs) [34].

Regenerative braking can be used to store and reuse generated electricity. The electricity is gained by dynamic braking; the electric motor is driven backward and should slow down the train. The electric motor can act as a generator with the aid of dynamic braking. Currently, most of the trains used to convert this dynamic braking energy and deploy resistance bridges. The resistance bridges heat up during this process. Usually, a cooling grid is arranged at the top of the locomotive for the braking resistors. The use of electronically controlled pneumatic (ECP) brakes is increasing, but energy savings can only be realized here, as long as all railcar brakes are connected. For regenerative braking to be used, an onboard rechargeable energy storage system (RESS) is required. Only with the RESS, the recovered kinetic energy can be stored and returned if needed. Normally, this kinetic energy converts to frictional heat and is thus lost. The recovered braking energy can be redirected back into the system or used for peak load requirements such as accelerating or uphill driving. Modern or newer electric train systems are able to save and reuse 10–20% of energy consumption with the aid of the regenerative braking function. For rail freight traffic ith numerous stops, this is particularly interesting in terms of energy effici cy [34].

Energy efficiency for high-speed rail

Freight railways deserve a lot of attention as they require about 90% of the energy of domestic rail transport. In 2011, the American Public Transportation Association (APTA) cited some UIC data in a report stating that the high-speed trains achieved about 106 mi (170 km) per kWh of energy. In comparison, planes only get 13 mi (21 km) and cars 34 mi (55 km) per kWh [35]. High-speed rails (HSR) and maglev (magnetic levitation) systems have many advantages over conventional rail, highway, and air, especially in terms of air quality and sustainability. CO_2 emissions from HSR operations are signifi antly lower (0.1–0.3 lb of CO_2/passenger-mile or 0.03–0.08 kg/passenger-km) than other modes of transport such as aircraft (0.6 lb./p-mi or 0.17 kg/pkm) or cars (0.5 lb/p-mi or 0.14 kg/pkm). An up-to-date life cycle analysis of HSR versus traditional rail, air, and highway modes found that system-wide comparisons of rail infrastructure construction and operation, high load and occupancy factors, maintenance and fuel, and clean electricity must be supplied with renewable energies (instead of, e.g., coal-fi ed power stations) [34].

Diesel multiple units (DMUs) and electric multiple units (EMUs)

Self-propelled diesel railcars can be diesel-electric, diesel-hydraulic, or diesel-mechanical units. These traction vehicles (including powered wheels) can accelerate much faster and have a shorter braking distance than locomotive trains and are thus also more energy-effici t than these. As long as the powered vehicles are connected by cable or radio link, they can be used in the decentralized power configur tion. Another form of self-propelled railcars are electric multiple units (EMUs). EMUs are individually powered by direct current (DC) from a third rail, but there is also the option of being powered by a vehicle pantograph in contact with the AC overhead line system (OCS). EMUs are more costly than DMUs but are more environmentally friendly and more energy-effici t and can achieve higher speeds [34].

Dual power hybrid locomotives

A **hybrid train** is a locomotive, railcar, or train which uses an onboard rechargeable energy storage system (RESS), placed between the power source (often a diesel engine) and the traction transmission system connected to the wheels. An example is the hybrid locomotive ALP-45DP with dual drive developed by Bombardier. It is designed as a diesel and electric locomotive. The hybrid locomotive can reach a speed of 160 km/h under diesel drive and up to 200 km/h under electric drive. For a higher effici cy, it can be switched from the diesel drive to the purely electrical operation with just a button push [34].

Efficient and ultraclean diesel-electric locomotives and repower kits

General Electric (GE) Evolution Series is a diesel-electric locomotive with 12-cylinder engine. It is currently considered the most fuel-effici t and technologically advanced. Th s product is viewed to be particularly environmentally friendly and currently has about 3700 locomotives in 10 different countries in use [34]. It has the advantage over other clean locomotives that no urea additive for selective catalytic reduction (SCR) is needed to reduce NO_x emissions. There are no expensive infrastructure upgrades to store and deliver urea for denitrifi ation. New locomotives are very expensive to buy, so a cost-effective option to achieve energy effici cy would be to retrofit existing locomotives with repower kits that reduce fuel consumption and emissions. Progress Rail Service (PRS) acquired in 2010 is a Caterpillar division, Electro-Motive Diesel (EMD), which has long been producing locomotives. The goal of the acquisition was to switch low-power, regional, and high-performance longrange mid-power locomotives to cleaner operation through repowering and achieving more effici t traction. The 710ECO Repower locomotives signifi antly reduce fuel consumption. Up to 25% of fuel and even 50% of lubricating oil can be saved. Th s is highlighted by the manufacturer as one of the most important advantages for railways, which are constantly confronted with rising fuel costs [34].

Distributed power management and control technologies for freight rail

Unlike the traditional push-pull configur tion, decentralized power supplies place locomotives in the middle and at the ends of the trains. Distributing the locomotive power along the train can achieve about 5% more energy effici cy (compared to the push-pull configur tion). Th s is increasingly being used by freight trains. Due to the power distribution, there is an increase in safety, as the trains are less susceptible to derailment [36]. The wear of wheels and tracks as well as the braking distance can be signifi antly shortened by the distributed power. To achieve a desired speed curve, distributed energy control and power management software is used. Various companies such as Canac and Wabtec offer solutions for the decentralized electricity market. Norfolk Southern saved nearly 30% on fuel with a combination of electronically controlled pneumatic (ECP) brakes that communicate with GE's Locotrol over the network with the LEADER train management and control system [34].

Alternative fuels for environmental sustainability

Hydrogen fuel for fuel cell hybrid locomotives

Hydrogen has been considered for use in rail transportation [37, 38]. Alternative fuels for locomotives are particularly important for environmental improvement in the rail sector. With the use of hydrogen fuel cells, a reduction of particulate matter pollution and greenhouse gases emitted into the atmosphere can be achieved. The dependence of the iron webs on fossil fuels is reduced to a minimum with the hydrogen fuel cell [34].

In the hydrogen hybrid locomotive, the following components are used:

- Batteries for driving electric traction motors, which are charged by a Ballard fuel cell stack with 240 kW
- Hydrogen tanks storing 70 kg of hydrogen at 350 bar on the roof
- Corresponding power electronics
- Battery ventilation systems

The batteries that drive the electric motor are charged by the fuel cell. To ensure sufficient traction between rails and wheels, the locomotive carries 900 kg of ballast. In the case that the temperatures of the batteries are too high, a pressure relief device can be activated. This process ventilates the batteries as well as the hydrogen fuel cells. With this model, the air pollution and the noise pollution at the stations are reduced. The problem with this variant is the limited range between the fueling and the hydrogen storage capacity [34].

Natural gas locomotives using liquefied natural gas (LNG)

Liquefi d natural gas (LNG) is an interesting alternative fuel for locomotives [39, 40]. Westport Innovations is working with Caterpillar to develop a natural gas fuel system for locomotives [34]. Th s project uses high-pressure direct injection technologies for combustion. The main objective was defi ed as the production of emission compliant long-haul locomotives with interchangeable tender vehicles. With this technology, 95% of the diesel fuel is replaced by natural gas and thus only 5% diesel fuel used for combustion to bring the locomotive to full capacity. Energy Conversions Inc. is working with Burlington Northern Santa Fe (BNSF) to develop a convertible engine with a dual-fuel system. Th s system uses low-pressure direct injection (LPDI) with no pump being required. The NOx emissions caused by premixed combustion are reduced. Th s system can save up to 1.1 million liters of diesel per year per locomotive, equivalent to a possible replacement of 92% [34]. According to BNSF, the economy and technology have been improved so much that natural gas in long-haul locomotives becomes operationally feasible [34].

Biofuels and blends with petrodiesel

Biofuels are derived from renewable and (in principle) non-exhaustive sources of energy. To produce biofuels, biological (plant or animal) materials are converted into liquid fuel composed of fatty acid methyl esters (FAME). Instead of fossil fuels, organic waste (e.g., waste cooking oil) can also be used for production [41, 42]. Biodiesel fuel is obtained from transesterifi ation of fatty acids. In this chemical process, glycerol is separated from fat or vegetable oil, and methanol is consumed. Biodiesel is made from a variety of products, such as animal fat, vegetable oil (rape seed, soy bean, palm oil, etc.), or recycled restaurant fat. Petroleum diesel can be blended with biodiesel to any percentage. In these biodiesel blends, the percentage of biodiesel is always clearly marked. For example, B10 contains 10% of biodiesel, with the remaining 90% being made from fossil sources. Pure biodiesel is known as B100. Blends containing more than 20% biodiesel require special handling or even modifi ations of the equipment. Biodiesel is biodegradable and nontoxic, reduces air pollutants, and provides better lubricity due to its viscosity. The high cetane number facilitates combustion in compression ignition engines [34].

For two recent reviews on biodiesel, see [43–45] for biodiesel in railway use. Bioethanol is more a fuel of choice for smaller (gasoline) engines. Other biofuels are, e.g., biobutanol and biomethanol [3, 46].

Airfreight

The steadily growing world trade is the reason for the rapid increase in air cargo volume in recent decades. Th s transport method offers many advantages such as speed, safety, and reliability. The short transport times over long distances are particularly attractive for goods with a high urgency and high value. Airfreight records the highest growth worldwide compared to other modes of transport [47]. Another advantage is the precisely planned organization in air traffic Flight plans are minutely adhered to under very high safety standards, thus ensuring a smooth supply chain. Compared to other transport methods, the transport costs, due to the high fuel consumption, are relatively high. Aircraft consume about 12 times more fuel than, for example, seagoing vessels [48] per tonne kilometer.

Based on the assessment basis for specific CO_2 emissions, air traffi is a signifi ant contributor to climate change. In most cases, energy consumption is related to transport performance, such as passenger kilometers or tonne kilometers. Th s includes the consumption from the departure terminal to the arrival terminal and therefore also the movements that take place on the ground. Between 1990 and 2011, freight transport services quadrupled in Germany, and on a global scale, a future annual growth rate of ~7% is expected on average [4, 47].

In December 2017, the International Air Transport Association (IATA) published updated data for the global airfreight market. It showed that demand (measured in freight tonne kilometers, FTK) increased by 5.9% compared to the previous year. Freight capacity, measured in available freight tonne kilometers (AFTKs), also increased by 3.7% compared to 2016 [49].

Alexandre de Juniac (IATA Director General and CEO) said: “Demand for air freight increased by 5.9% in October. And tightening supply conditions in the fourth quarter should be the air cargo industry delivering its strongest operational and fi ancial performance since the post-global fi ancial crisis rebound in 2010” [49].

In the Asia-Pacific region, airlines increased their cargo volumes by 4.4% and capacity by 3.9%. Freight demand exceeds the record high reached in 2010 by around 3%.

Airlines in North America recorded an increase in cargo volume of 6.6% in 2017 compared to 2016. The increase in capacity was 3.8%. In recent years, the market for inbound freight transport has increased due to the strength of the US economy and the US dollar.

In Europe, the 5-year average of 4.9% was exceeded, and freight demand rose by a total of 6.4%. Capacity grew by 2.5%. Compared to other continents, European export orders have been rising fastest for more than 7 years [49].

Middle Eastern carriers’ freight volumes increased 4.6%, and capacity increased 3.4% in 2017 [49].

In the last half of 2017, seasonally adjusted international freight volumes continued to rise at a rate of 8–10%. Airlines in Latin America, like all other major regions, posted positive growth in freight demand (7.2%) and capacity (4.4%). By far the largest increase over the previous year was seen by African carriers. Freight demand rose by 30.3% and capacity by as much as 9.2% [49].

Decarbonization attempts in aviation concern passenger and freight transport alike. Engine improvements have a very strong leverage on energy effici cy. There is a trade-off between NOx emissions and turbine energy effici cy [50].

Solar energy systems: solar kerosene

For many years countless research activities have been dealing with the topic of solar energy and where it can be used. The EU Commission announced in 2014 that an experiment had succeeded in producing kerosene with the help of sunlight [51]. In the process, synthesis gas is generated under the action of sunlight, which consists of hydrogen (H_2) and carbon monoxide (CO). Andreas Sizmann from the Bauhaus Luftfahrt (participant in the research project) explained two major advantages of this method. First, the harmful climate gas CO_2 would be used and not fossil hydrocarbons such as oil. Although the kerosene produced in this way will also release CO_2 through combustion, CO_2 can be obtained directly from the air over the long term. Therefore, the process is on the whole potentially CO_2 neutral, according to Sizmann. Second, the energy for the entire process is generated from solar energy. The process is very effici t and does not compete with food production as opposed to the production of other (mainly fi st-generation) biofuels [52].

Electric motors: environmentally friendly flying

The use of electric motors is already well advanced in parts of the transport sector. Soon, electric flying should become possible. In this regard, Siemens and Airbus announced a development cooperation in 2016 in which hybrid technology is used. In the presented test aircraft, the jet engine was replaced by a 2 megawatt electric motor (produced by Siemens), which drives the large air impellers. The 2 megawatt electric motor is only 30 centimeters long and weighs 175 kilos. To get power from the electric engine, several steps are necessary. With a gas turbine burning kerosene, an electric generator is powered, which feeds the power into a 2 tonne lithium-ion battery. Finally, the lithium-ion battery supplies the built-in electric motor. Since starting up an aircraft requires a great deal of energy, the lifting can be supported by generator and battery. During the descent, the engine blades, which work like small windmills, can be used to generate electricity. Th s principle is similar to that of electric cars or locomotives, which carry power back into the battery while braking. The representatives of this project are of the opinion that with their concept they can reduce the consumption of kerosene by double-digit percentages compared to conventional jet engines. Flying would therefore also become more environmentally friendly and more quiet [53, 54]. Electric power for a two-seat aircraft is discussed in [55]. The more electric aircraft (MEA) concept is discussed in [56]. Light pureelectric and hybrid-electric aircraft are presented in [57]. The MEA concept essentially aims at replacing conventional non-electric power (pneumatic, hydraulic, and mechanical) by electric power to drive aircraft subsystems more effici tly. An all-electric 180-passenger commercial aircraft s discussed in [58].

Aerodynamics: winglets and riblets

In aviation, aerodynamics focuses on two main forces: lift and drag. The power of lifting makes an airplane fly. Th s is caused by the uneven pressure on a wing's top and bottom. The drag represents the resistance that arises during movement through the airfl w. Due to the high pressure under the wings, air fl ws over the wing tips upward and rolls off in the form of a vortex. Th s vortex is also called induced drag and can be so strong that it disturbs other planes. Wake turbulence can become a safety concern particularly for small aircraft. nduced air drag degrades performance and reduces the range and speed of the aircraft [59]

Winglets are more than just a striking and aesthetic design feature; they are among the most visible fuel-saving and performance-enhancing technologies in aviation introduced in recent years. According to Whitecomb, winglets can reduce induced drag by about 20% and improve carrying capacity by 6–9%. The design of the winglets can be very different. Aviation Partners Boeing (APB) has developed a special form: the Blended Winglet. The Blended Winglet's design fuses the wing into a smooth upward curve. Other winglets are shaped more like a fold or kink. Th ough this smooth transition, optimal effici cy can be achieved [59].

Riblets are micro- and nanostructured surface structures that cause drag reduction. Th s technology comes from the fi ld of bionics, which works by transferring phenomena from nature to technology. Riblets resemble the skin of a shark and are characterized by fi e grooves on the surface. The so-called shark skin effect causes a reduction of the friction resistance of up to 8% compared to aircraft ithout this coating [60].

In the research project FAMOS (management system for the automated application of multifunctional surface structures) of Lufthansa Technik, Airbus Operations GmbH, BWM GmbH, and the Fraunhofer Institute for Production Technology, it has been possible to develop an automatic guidance system for the application of riblets to the outer shell of the test aircraft [61]. Tests from this research project have shown that riblets, despite minor wear of the microstructures, signifi antly reduce the frictional drag in the air. For riblets, lacquer is fi st applied to a UV-transparent mold or matrix. Th s matrix contains the negative impression of the riblet shape. The resulting negative mold is then pressed into fresh paint and thus cured with UV light. After removing the negative mold, the positive of the sharkskin structure stops at the surface. The application of the riblets is possible on any aircraft models; they are attached in the form of strip tracks on the surfaces parallel to the fl w direction. In the laboratory of the project FAMOS, the longevity as well as the effici cy of the sharkskin structure was confi med. Depending on the area applied, airlines can use this technology to save about 1.5% of fuel [61].

Composite materials to optimize fuel consumption and CO_2 emissions

The processing of composite materials is becoming increasingly important for aircraft construction. Even though planes are themselves tonnage heavy, every single kilogram counts. The manufacturing and processing costs of carbon fiber-reinforced plastics (CFRP) in aircraft far exceed the costs of traditional metal construction. In the long term, however, the cost advantage outweighs due to the low weight and the resulting reduced fuel consumption. Nowadays, fuel consumption is a top priority for airlines because less fuel means less CO_2 emissions and lower operating costs. Thus, something good is done for the environment while saving money, too. For many years CFRP has been installed on models such as the Airbus A380 (28%) or the Airbus A350 XWB (53%). Predecessors, such as the A330, weigh almost 10 tonnes more and consume more fuel than the Airbus A350 XWB with comparable payload capacity and range [62, 63].

CFRP consist of hair-thin layers of carbon fibers, which are embedded in a resin matrix (thermoset). The material scores with a very high specific strength and low weight. Mechanically, this composite material is extremely difficult to deal with,

so millimeter-thin CFRP tapes have to be stacked on top of each other for the outer hull of the skin in a daylong process and then baked together under pressure and heat. Due to the extreme hardness of the material, particularly high-quality and expensive cutout drills and cutters (e.g., for external connections, doors, windows, and holes for rivets) must be used. Because of the high abrasion when drilling, even modern tools with diamond-like coating last on average only half as long as tools in metalworking [62].

Aircraft engines: current technology and energy-efficient developments

Aircraft engines must be reliable and effici t. The technology behind them is explained quickly and easily. Engines work in a similar way as rockets: The intake air is compressed and fuel is injected. The combustion of the fuel creates an exhaust gas jet, which emerges at the back. The exhaust jet drives the actual turbine (a wheel with blades). The turbine fi ally generates the drive for the compressor at the engine entrance. The compressor increases the pressure of the air and consists of several stages. Each of these stages includes a rotor and a stator wheel. The turbine part is also constructed like that. Depending on the engine, between 8 and 14 stages are used today. Particularly modern engines achieve compressions of 45 times the input pressure. The developments in aircraft engines initially focused on sending part of the air through the compressor and the combustion chamber, rather than around the engine. The fi st stage of the compressor, also known as a fan, works like a giant blower. The fan accelerates this circulating mantle air. The so-called turbofan engines are the current state of the art [64].

The shroud fl w ideally requires a relatively low speed for the large fan and high speed in the high-pressure range. Th s created the two-shaft engines. The axles of these engines can rotate counter-wise. One of them is the slow-speed low-pressure shaft, which is driven by the rear turbine stages just before the exhaust outlet. At the same time, the fi st compressor stages are rotated. The other one is the very fast-rotating high-pressure shaft. The high-pressure shaft is operated by the turbine stages behind the combustion chamber and thus moves the high-pressure part of the compressor.

Optimization of the engine concept has been in progress for many years. First and foremost, the approach is followed to change the amount of air that has passed through. The difference in speed should not be too big between thrust and airspeed. Ideally, a very large amount of air is pushed back very slowly from the engine. Another approach for increasing effici cy is the turbine including the combustion chamber. The hotter the combustion is, the more effici t the process becomes. Here, the materials are pushed to their limits. The fi st stage of the turbine is under most stress because it gets the full heat of the combustion chamber. Other developments are heading back in the direction of the classic propeller. Ideas in this area run under the slogan "open rotor concept." However, the mounting size, which makes mounting on the wing difficult, as well as the noise, proves to be problematic. Aircraft could look completely different in the future, for example, with a huge propeller engine on the roof of the fuselage.

Continuous descent operations: CDO

Continuous descent arrival (CDA) is an aircraft operating technique designed to reduce aircraft noise, fuel consumption, and emissions. In this method, an incoming aircraft sinks with minimal engine performance and largely avoids horizontal fli ht phases. Ideally, this happens at idle. Without the use of CDA, an airplane goes down step-by-step. When performing CDA, the aircraft lingers high up in the air for extended periods of time, operating at a low engine thrust. Th s causes a reduction

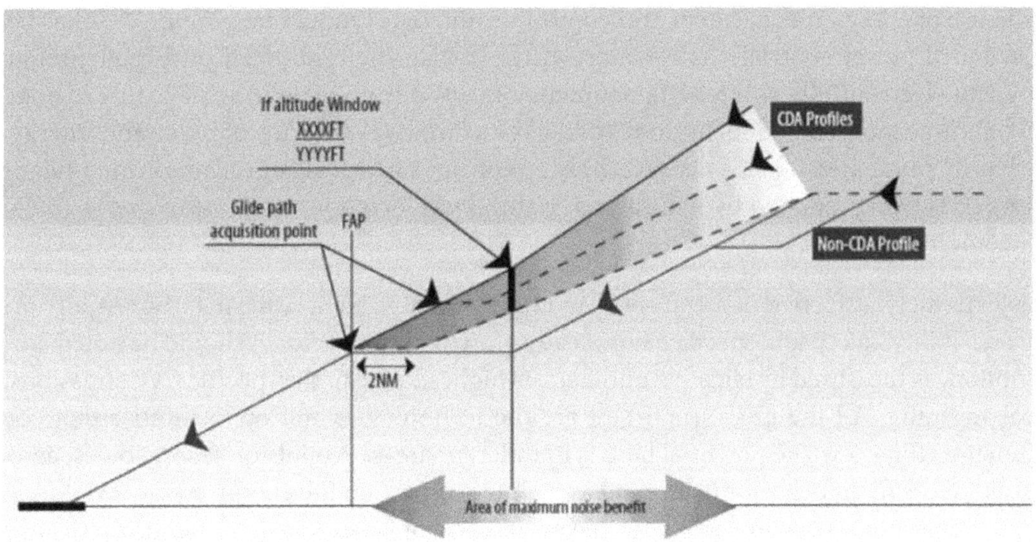

Figure 7.
Visual representation of CDA [65]. NM = nautical mile; FAP = final approach point; FT = foot.

in fuel consumption, emissions, and noise during the descent profile. The most optimum CDA starts at the top of the descent and ends as soon as the plane starts the last approach and follows the glide path to the runway [54] (compare **Figure 7**).

Conclusion

To sum up, energy effici cy management should always be used and aspired to because of the overwhelming economic and ecologic benefits in the transport sector. One of the most important advantages is, above all, the potential for saving fuel, since low fuel consumption leads to cost savings and consequently to reduced CO2 emissions. The application of energy-effici t technologies and methods therefore not only has economic aspects but is also good for the environment. Th s chapter shows that there are already several propulsion technologies and developments in the areas of rail freight and airfreight, but they are not yet completely revolutionized and provide room for further improvements.

Rail freight currently has mainly internal combustion engines and electric motors in use. In particular, the invention of electrically powered railroads has been a signifi ant advance in improving energy effici cy. By using modern drive technologies and methods in rail freight transport, many potential savings can be achieved. In the fi ld of electric locomotives, regenerative braking can be used, for example, with rechargeable energy storage systems (RESS). Th s allows the current generated during dynamic braking to be stored and reused, and it offers a potential of 10–20% of energy consumption to save. Other powertrain technologies such as the dual power hybrid locomotive as well as diesel multiple units (DMUs) and electric multiple units (EMUs) are particularly effective and effici t. The dual power hybrid locomotive convinces with the advantage that the diesel engine can be converted to a purely electric drive with just a single push of a button. Although EMUs prove to be a high fi ancial burden, they are considered to be very environmentally friendly and energy-effici t. The most fuel-effici t and advanced engine in this regard is the highly clean diesel-electric locomotive with repower kits. Compared to other locomotives, this model has the advantage of not requiring urea additives to reduce NO_x emissions. Since the purchase of a completely new diesel-electric locomotive is extremely expensive, repower kits are a cost-effective option that can also reduce fuel consumption and emissions. Th s can save up to 25% on fuel and about 50% on lubricating oil. To optimize energy effici cy, the use of energy management and control technologies should be promoted. It has been proven that the distributed power controller is 5% more effici t than the traditional push-pull configur tion. In conjunction with electronically controlled pneumatic brakes, it is possible to achieve fuel savings of almost 30%. Rail freight transport should focus on natural gas locomotives in terms of renewable energy. With this technology hardly more diesel fuel is needed. Only 5% of the diesel fuel is needed to reach full power, with the remaining 95% being replaced by natural gas. Natural gas locomotives can save over 1 million liters of diesel per locomotive per year.

Airfreight is extremely attractive as a transport method. The aircraft as a means of transport brings many benefits. The reliability, safety, and speed ensure steady growth in this sector. It should be noted, however, that fuel consumption, as measured by effective numbers, is highest. Due to the specific CO_2 emissions, airfreight drives global warming. All the more important are the technologies and opportunities that contribute to improving energy effici cy. The potential here is in different areas. Propulsion technologies, aerodynamics, composites, as well as fli ht behavior itself can have a huge impact on the energy effici cy of freighters. The solar energy systems announced by the EU for the production of "solar kerosene" act primarily as an optimal way to supply clean fuel to aircraft, but in this process, CO_2 is also produced by combustion. Electric motors in aviation are currently still in the development phase but could become very relevant in terms of energy effici cy in the future. The developments in the fi ld of aerodynamics have proven to be effective and operational for many years. With the help of winglets and riblets, air resistance can be reduced by 20% and load capacity increased by almost 9%. Th s technology can be applied to any aircraft and offers around 1.5% fuel savings.

In aircraft construction, the processing of composite materials can result in optimized fuel consumption and a reduction in CO_2 emissions. A disadvantage of carbon fiber-reinforced plastics is the high manufacturing and processing costs. In the long term, however, these investment costs are off et by the respective savings. Composites are particularly interesting because of their high specific strength. Changing the fli ht behavior can result in a reduction in fuel consumption, emissions, and aircraft noise. Th s aircraft operating technique is called continuous descent arrival. The state of the art is currently the turbofan engine. The developments have shown that ideas in the direction of the classic propeller cannot be ruled out.

Acknowledgements

The authors want to thank Hannes Grünbichler for the support in manuscript proofreading.

Conflict of interest

There is no confli t of interest.

List of abbreviations

AC alternating current

ACEA European Automobile Manufacturers Association

AFTK available freight tonne kilometer

APB Aviation Partners Boeing

APTA American Public Transportation Association

ASME American Society of Mechanical Engineers

BNSF Burlington Northern Santa Fe

CDA continuous descent arrival

CDO continuous descent operation

CFRP carbon fiber-reinforced plastic

CH_4 methane

CO_2 carbon dioxide

CO_{2eq} CO2 equivalents

comb. combustible

DC direct current

DMU diesel multiple unit

ECP	electronically controlled pneumatic
EMD	electro-motive diesel
EMU	electric multiple unit
EPA	Environmental Protection Agency
EU	European Union
EU ETS	European Emissions Trading System
FRA	Federal Railroad Administration
FTK	freight tonne kilometer
gas	natural gas
GDP	gross domestic product
GE	general electric
GHG	greenhouse gas
GJ	gigajoule (109 joules)
GWP	global warming potential
H_2	hydrogen
HFC	hydrofluorocarbon
HSR	high-speed rail
HVAC	heat, ventilation, and air conditioning
IATA	International Air Transport Association
IEA	International Energy Agency
LED	light-emitting diode
LNG	liquefi d natural gas
LPDI	low-pressure direct injection
LPG	liquefi d petroleum gases
m^2	square meter
MJ	megajoule (106 joules)
N_2O	nitrous oxide
NAZCA	non-state actor for climate action

NF_3	nitrogen trifluoride
PFC	perfluorocarbon
PJ	petajoule (1015 joules)
pkm	person kilometers
PRS	progress rail service
RESS	rechargeable energy storage system
SCR	selective catalytic reduction
SF_6	sulfur hexafluoride
tkm	tonne kilometers
TREMOD	transport emission model
UNFCCC	United Nations Framework Convention on Climate Change
VA	value added
WEO	World Energy Outlook
WESS	wayside energy storage system

Author details

Corinna Cermak, Erich Markl* and Maximilian Lackner
University of Applied Sciences FH Technikum Wien, Vienna, Austria

*Address all correspondence to: erich.markl@technikum-wien.at

References

[1] https://www.oxfam.de/unserearbeit/themen/folgen-klimawandels [Accessed: February 23, 2019]

[2] Palm J. Energy Effici cy. London, UK: IntechOpen; 2010. ISBN 978- 953-307-137-4. Available from: https://www.intechopen.com/books/ Energy-Effici cy

[3] Chen W-Y, Suzuki T, Lackner M. Handbook of Climate Change Mitigation and Adaptation. Switzerland: Springer International Publishing; 2017. ISBN 978-3-319-14410-8

[4] Sauer A, Bauernhansl T, editors. Energieeffizien in Deutschland– eine Metastudie. 2nd ed. Zurich, Switzerland: Springer, Vieweg; 2016. ISBN: 978-3-662-48882-9

[5] Mathy S, Menanteau P, Criqui P. After the Paris agreement: Measuring the global decarbonization wedges from national energy scenarios. Ecological Economics. 2018;**150**:273-289

[6] https://www.bmwfw. gv.at/EnergieUndBergbau/ Energieversorgung/Documents/ de_EE%20RL_2009_28_EG.pdf; page:L 140/16 (1) [Accessed: February 23, 2019]

[7] http://klimaenergie-frm.de/ Klima-Energie-Wissen/Politische- Rahmenbedingungen/index.php?mNavID=2617.58&sNavID=2617.58&La=1 [Accessed: February 23, 2019]

[8] http://klimaenergie-frm.de/ Klima-EnergieWissen/Politische- Rahmenbedingungen/index.php?mNavID=2617.58&sNavID=2617.58&La=1 [Accessed: February 23, 2019]

[9] http://klimaenergie-frm. de/?NavID=2617.58&La=1 [Accessed: February 23, 2019]

[10] IEA. World Energy Outlook 2014. Paris, France; 2014. ISBN 978-92-64-20804-9. Available from: http://www.iea.org/W/bookshop/477- World_Energy_Outlook_2014 [Accessed: February 23, 2019]

[11] http://www.iea.org/topics/ energyeffici cy/ [Accessed: February 23, 2019]

[12] Murray G, Patterson. What is energy effici cy? Concepts, indicators and methodological issues. Energy Policy. 1996;**24**(5):377-390

[13] http://www.iea.org/publications/ freepublications/publication/Energy Effici cyIndicatorsHighlights_2016. pdf [Accessed: February 23, 2019]

[14] http://www.acea.be/statistics/ tag/category/share-of-diesel-in-new- passenger-cars [Accessed: February 23, 2019]

[15] Xu Y, Kang H, Gong J, Zhang S, Li X. A study on the combustion strategy of gasoline/diesel dual-fuel engine. Fuel. 2018;**225**(1):426-435

[16] Wilberforce T, El-Hassan Z, Khatib FN, Al Makky A, Olabi AG. Developments of electric cars and fuel cell hydrogen electric cars. International Journal of Hydrogen Energy. 2017;**42**(40):25695-25734

[17] Danielis R, Giansoldati M, Rotaris L. A probabilistic total cost of ownership model to evaluate the current and future prospects of electric cars uptake in Italy. Energy Policy. 2018;**119**:268-281

[18] https://www.ipc.be/en/knowledge- centre/e-commerce/ Articles/global- ecommerce-figu es-2017 [Accessed: February 23, 2019]

[19] Hamamcioglu C, Oguztimur S. The comparison of basic transportation indicators and freight villages' locations between Germany and Turkey. In: 55th Congress of the European Regional Science Association: "World Renaissance: Changing roles for people and places"; Lisbon, Portugal. 2015

[20] https://www.destatis.de/DE /Publikationen/Thematisch/Umweltoe konomischeGesamtrechnungen/ VerkehrundUmwelt/UGRTransport leistungenEnergieverbrauch 5850010149004.pdf? blob=publication File [Accessed: February 23, 2019]

[21] Umweltbundesamt Editor. Entwicklung der Treibhausgasemissionen in Deutschland nach Sektoren. Germany: Dessau-Roßlau. 2014. https://www.umweltbundesamt. de/ sites/default/files/medien/376/ bilder/dateien/entwicklung_ der_treibhausgasemissionen_ in_deutschland_nach_ sektoren_1990bis2012_pi-2014-03_ anlage.pdf [Accessed May 16, 2019]

[22] Bockris J. Global Warming. London, UK: IntechOpen; 2010. Available from: https://www.intechopen.com/books/ global-warming/global-warming. DOI: 10.5772/10290

[23] https://www.toptarif.de/strom/ wissen/co2-emissionen/ [Accessed: February 23, 2019]

[24] IPCC. Special Report, Global Warming of 1.5°C. Geneva, Switzerland; ISBN 978-92-9169-151-7. Available from:, https://www.ipcc.ch/ sr15/ [Accessed: February 23, 2019]

[25] https://unfccc.int/process-and- meetings/the-paris-agreement/the- paris-agreement [Accessed: February 23, 2019]

[26] https://unfccc.int/process/the- kyoto-protocol [Accessed: February 23, 2019]

[27] https://de.statista.com/statistik/ daten/studie/179260/ umfrage/die-zehn-groessten-c02-emittenten- weltweit/ [Accessed: February 23, 2019]

[28] https://volker-quaschning.de/ datserv/CO2/index.php [Accessed: February 23, 2019]

[29] https://www.ifeu.de/en/methods/ models/tremod/ [Accessed: February 23, 2019]

[30] https://www.ifeu.de/methoden/ modelle/tremod/2017-05-27_ tremod-co2-emissionen/ [Accessed: February 23, 2019]

[31] https://www.sci.de/fileadmin/ user_upload/presse/pdf_ downloads/170308_Pressemitteilung_ Schienenverkehrsmarkt.pdf [Accessed: February 23, 2019]

[32] Kipp L. Energieeffizien in der Logistik. 1st ed. Munich, Germany: GRIN Verlag. ISBN (eBook): 9783656678229

[33] http://deacademic.com/dic.nsf/ dewiki/561583 [Accessed: February 23, 2019]

[34] www.kpesic.com/sites/default/files/ DOT-VNTSC-FRA-13-02.pdf [Accessed: February 23, 2019]

[35] www.apta.com/resources/ reportsandpublications/documents/ HSRPub_fi al.pdf [Accessed: February 23, 2019]

[36] Liu X, Rapik Saat M, Barkan CPL. Freight-train derailment rates for railroad safety and risk analysis. Accident Analysis & Prevention. 2017;**98**:1-9

[37] Ahmed A, Al-Amin AQ, Ambrose AF, Saidur R, fuel H, system t. A sustainable and environmental future. International Journal of Hydrogen Energy. 2016;**41**(3):1369-1380

[38] Fragiacomo P, Francesco P. Energy performance of a fuel cell hybrid system for rail vehicle propulsion. Energy Procedia. 2017;**126**:1051-1058

[39] Dincer I, Zamfi escu C. A review of novel energy options for clean rail applications. Journal of Natural Gas Science and Engineering. 2016;**28**:461-478

[40] Osorio-Tejada JL, Llera-Sastresa E, Scarpellini S. Liquefi d natural gas: Could it be a reliable option for road freight transport in the EU? Renewable and Sustainable Energy Reviews. 2017;**71**:785-795

[41] Abed KA, El Morsi AK, Sayed MM, El Shaib AA, Gad MS. Effect of waste cooking-oil biodiesel on performance and exhaust emissions of a diesel engine. Egyptian Journal of Petroleum. 2018;**27**(4):985-989

[42] Sahar SS, Iqbal J, Ullah I, Iqbal M. Biodiesel production from waste cooking oil: An effici t technique to convert waste into biodiesel. Sustainable Cities and Society. 2018;**41**:220-226

[43] Gebremariam SN, Marchetti JM. Economics of biodiesel production: Review. Energy Conversion and Management. 2018;**168**(15):74-84

[44] Ambat I, Srivastava V, Sillanpää M. Recent advancement in biodiesel production methodologies using various feedstock: A review. Renewable and Sustainable Energy Reviews. 2018;**90**:356-369

[45] Vaiciunas G, Lingaitis LP. Biodiesel features in the railway transport. In: Bernardes MADS, editor. Economic Effects of Biofuel Production. London, UK: IntechOpen; 2011. Available from: https://www.intechopen.com/books/ economic-effects-of-biofuel-production/ biodiesel-features-in-the-railway-transport

[46] Brito Cruz CH, Souza GM, Barbosa Cortez LA. Chapter 11: Biofuels for Transport. In: Salles-Filho SLM, Cortez LAB, da Silveira JMFJ, Trindade S, Fonseca MGD, editors. Global Bioethanol: Evolution, Risks, and Uncertainties. Future Energy. 2nd ed. London, UK: Academic Press; 2011, 2014. pp. 215-244. ISBN: 978-0-12-803141-4

[47] http://www.iata.org/pressroom/pr/ Pages/2018-01-31-01.aspx [Accessed: February 23, 2019]

[48] http://www.logistikbranche.net/ dossier/vorteil-nachteil-luftfracht.html [Accessed: February 23, 2019]

[49] http://www.iata.org/pressroom/pr/ Pages/2017-12-01-01.aspx [Accessed: February 23, 2019]

[50] Kyprianidis KG, Dahlquist E. On the trade-off between aviation NOx and energy effici cy. Applied Energy. 2017;**185**(Part 2):1506-1516

[51] http://europa.eu/rapid/press- release_IP-14-481_en.htm [Accessed: February 23, 2019]

[52] http://www.handelsblatt.com/ technik/das-technologie-update/ energie/nachhaltige-kraft toffe- flie en-flugzeuge-bald-mit- solarkerosin/9819784.html [Accessed: February 23, 2019]

[53] https://www.welt.de/wirtschaft/ article171058947/Mit-diesem- Triebwerk-startet-die-Aera-grosser-E- Flugzeuge.html [Accessed: February 23, 2019]

[54] https://www.siemens.com/press/ de/feature/2015/corporate/2015-03- electromotor.php?content[Z]=Corp [Accessed: February 23, 2019]

[55] Xiang S, Liu Y-Q , Tong G, Zhao W-P, Li Y-D. An improved propeller design method for the electric aircraft. Aerospace Science and Technology. 2018;**78**:488-493

[56] Yang Y, Gao Z. Power optimization of the environmental control system for the civil more electric aircraft. Energy. 2019;**172**(1):196-206

[57] Riboldi CED, Gualdoni F, Trainelli L. Preliminary weight sizing of light pure-electric and hybrid-electric aircraft. Transportation Research Procedia. 2018;**29**:376-389

[58] Gnadt AR, Speth RL, Sabnis JS, Barrett SRH. Technical and environmental assessment of all- electric 180-passenger commercial aircraft. Progress in Aerospace Sciences. 2019;**105**:1-30

[59] https://spinoff.nasa.gov/ Spinoff2010/t_5. tml [Accessed: February 23, 2019]

[60] http://www.bionicsurface.com/ riblet-oberflaechen/ [Accessed: February 23, 2019]

[61] https://www.lufthansa-technik. com/de/famos [Accessed: February 23, 2019]

[62] http://www.handelsblatt.com/ technik/das-technologie-update/ frage-der-woche/luftfahrt-warum- werden-moderne-flugzeuge-aus- kohlefaser-hergestellt/10886940.html [Accessed: February 23, 2019]

[63] http://www.plastverarbeiter. de/15818/airbus-a350-xwb-besteht-zu- 53-aus-kohlefaser-verbundwerkstoffen/ [Accessed: February 23, 2019]

[64] Dik A, Bitén N, Zaccaria V, Aslanidou I, Kyprianidis KG. Conceptual design of a 3-shaft turbofan engine with reduced fuel consumption for 2025. Energy Procedia. 2017;**142**:1728-1735

[65] www.skybrary.aero/index.php/ File:Cpz20r26oo.jpg [Accessed: February 23, 2019]

4

Analytical Assessment of Effective Maintenance Operations on At-Grade Unsignalized Intersections

Francesca Russo, Salvatore Antonio Biancardo and Rosa Veropalumbo

Abstract

Th s chapter describes a methodological structure to support and improve the decision-making process for redesigning the geometric configur tions of substandard sites and thus reduce crash risk factors on at-grade three-leg and four-leg intersections with stop-control on minor roads and single-lane roundabouts belonging to a two-lane rural road network located in Southern Italy. Starting from an initial evaluation of the risk level at each investigated site and adopting a procedure developed by the Italian National Research Council based on an estimated crash rate level, a more precise hierarchy of intersections with "black" rankings was developed. In addition, new geometric configur tions for the most hazardous sites were suggested based on a statistical comparison in terms of safety and Level of Service (LoS). The effectiveness of the strategies was validated by computing the expected LoS and safety by adopting an empirical Bayesian analysis and performance functions centered on a revised Highway Safety Manual procedure refl cting the context of the study.

Keywords: risk assessment, reducing crash risk levels, level of service

Literature review

The key role played by transportation networks in social well-being and safeguarding the world economy means priorities must be established to maintain an adequate level of service and functionality and adequately managing existing weak areas as well as possible hazardous events: a thorough examination of activities within the system, potential risks for users, and the careful management of the planning phase of controls and maintenance operations can help reduce, if not even prevent, failures in the system that may compromise good operation and endanger health, safety, and the environment.

Dickey and Santos [1] identifi d the response time of emergency services during hazardous events in the transportation system—one of the fundamental actions in restoring disrupted infrastructures—and in guaranteeing essential levels of service and safety to users.

Freiria et al. [2] considered the road transport system as one of the most critical infrastructures in hazard situations performing an LRSRM model (Local Regional Scale Risk Model) to identify the most signifi ant roads from the multiscale perspective, which should guarantee better operability of the sites and help allocate local resources better during hazardous events.

European Directive 2008/96/EC [3] on road safety stressed the central role of risk analysis and management as activities that help ensure the good functioning of a road network, defini g road infrastructures as the third pillar of safety policy.

Many scholars [4–6] focusing on the road hotspots identifi d in the light of European Directive objectives suggested calculating crash frequencies and crash rates to rank "black" sites, while others suggested adopting the empirical Bayes (EB) approach and the full Bayes (FB) approach in combination with the previous measures.

The main reason for using a Bayesian approach is to force the analyst to look at historical data sets or to canvass expert knowledge to determine what is known about the parameters and processes [7–9]. The key difference between Bayesian statistical inference and frequentist statistical methods concerns the nature of the unknown parameters. In the frequentist framework, a parameter of interest is assumed to be unknown, but fi ed. In the Bayesian view of subjective probability, all unknown parameters are treated as uncertain and therefore should be described by a probability distribution. Replication is an important and indispensable tool [10], and Bayesian methods fit within this framework because background knowledge is integrated into the statistical model.

Xie et al. [11] worked out a procedure to identify hotspots in a road network, also investigating different contributing factors to road pedestrian safety such as vehicle volumes, road networks, land use, demographic and economic features, and the social media. The researchers identified potential "black" sites by estimating crash costs, considered an accurate safety measure well able to reflect injury severity levels.

Goals definition

Analysis procedure presented here focuses on intersections: crossing and turning maneuvers create opportunities for vehicle-vehicle, vehicle-pedestrian, and

vehicle-bicycle confli ts that may also result in traffi crashes. Certainly, human error is a contributing factor to road crashes; however, in addition to driver behavior, road engineering and design measures can also make intersections safer.

The work phases are shown in **Figure 1.**

In particular, the research steps are summarized as follows:

- Evaluating a fi st measure of exposure to crash risk by using a procedure developed by the Italian National Research Council [12], shown in detail in Section 3.1 and in Figure 3, it is useful in ranking black intersections.
- Computing LoS, determined by ascertaining control delay at each maneuver and estimating crash costs. Delays were assessed by revising specific analytical HCM 2016 [13] models on the basis of fi ld measurements (see Section 3.2 for details). The crash cost estimates were obtained from the mean values of the costs for injuries to people and damage to vehicles made available by the Italian Ministry of Infrastructures and Transport.
- Identifying hotspots: working in accordance with European Directive 2008/96/EC on road safety, the most dangerous intersections where high crash rate, high crash cost, and low-medium LoS were observed (hotspots) were analyzed in greater depth.

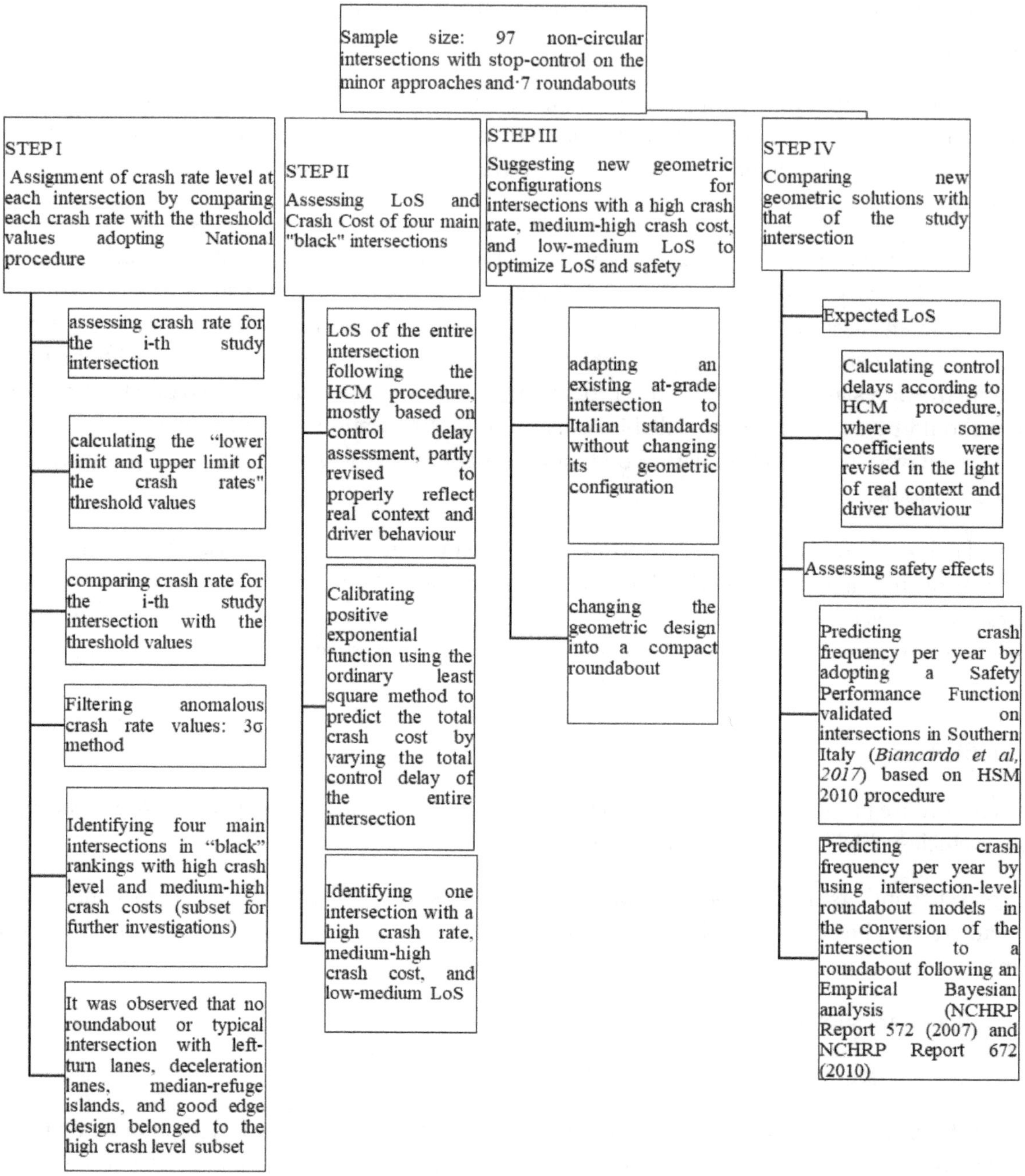

Figure 1.
Methodological approach to plan safer strategies at intersections.

- Identifying driver risk factors at hotspots by focusing on the mismatch between geometric design for each intersection and the requirements of the Italian Design Standard (Norme funzionali e geometriche per la costruzione delle intersezioni stradali, GU n. 70 del 24-7-2006) [14]; poor matching of the real configur tion with the design requirements is refl cted in greater consequent exposure to crash events and a lower level of service.

- Managing risk levels at hotspots by hypothesizing structural adjustments keeping traffi features and environmental conditions constant. Two adjustments were proposed:

- ◦ Adjusting the current geometric design to the Italian Design Standard without changing the configur tion [15, 16].
- ◦ Modifying the configur tion into a roundabout to achieve benefits in terms of a reduction of confli t points, vehicle speed reduction around the central island, and pedestrian safety [5, 17, 18].

- Assessing the effectiveness of risk management: comparing before and after configur tions of the hotspots by calculating the expected LoS and the expected safety effects in terms of crash frequency.

In greater detail, the expected LoS effects were calculated following HCM2016 procedure, but revising, in the light of measurements obtained at study sites; on the other hand, the expected computation of safety effects was performed by adopting the Safety Performance Function (SPF) introduced in [16] according to Highway Safety Manual (HSM) 2010 [17] procedure but revised in the light of study carried out in Southern Italy to which the intersections investigated here belong. Calculation of the expected safety effects obtained from converting the intersections to roundabouts was performed by (a) adopting the analytical models proposed in Rodegerdts et al. [18, 19], whose calibration conditions fit the study context presented here and (b) by using EB analysis to quantify the positive advance of intervention, a common statistical practice in the scientific l terature.

Th s book chapter is organized as follows: Section 2 focuses on data collection, while Section 3 focuses on data analysis for evaluating measures that refl ct the exposure of sites to crash risk; the results of the case study are displayed and discussed in Section 4.

Data collection

The crash data used involved 104 intersections belonging to two-lane rural roads in Southern Italy located in a flat area with a vertical grade of less than 5%, that is, 97 non-circular intersections before the Italian Road Design Standard [20] became law and seven single-lane roundabouts were built.

The road surface of the intersection area is paved, and neither space for pedestrian and bicycle use nor space to park vehicles exists by the roadside. In particular, 10% of the typical intersections have exclusive left- urn lanes, a suffici t sight distance for drivers at a stop-controlled or yield-controlled approaches, an acceptable entry and exit edge radius, and a median-refuge island along the legs. Almost all intersections have right-turn lanes on major-road approaches. The mean value of the approaching lane width and departing lane width is 2.70 m, and the average value of the entry and exit radius is 8 m; the average speed on the road segments belonging to the major roads approaching the intersections is around 70 km/h on a road Section 150 m from the intersection area, while on the minor road, it is approximately 45 km/h at the control section and 150 m from the intersection area.

All the single-lane roundabouts analyzed here are of the modern type. There are three conventional roundabouts with an inscribed circle diameter of between 40 and 50 m, three compact roundabouts with an inscribed circle diameter of between 25 and 40 m, and one mini-roundabout with an inscribed circle diameter of between 14 and 25 m. All the roundabouts have one entry lane for each approach as well as for the exit lanes of all the departures with an average entry and exit width of 3.00 m; the circulatory roadway has no lane markings, and the average width is 6.00 m. The circular central island is not practicable, and the average width is 4.00 m. The length of the splitter islands is almost 3.50 m, and the average width is 1.50 m. The average entry radius is 15 m with an exit radius of 25 m. The distance between an entry lane and the fi st exit lane for the next leg is at least 10 m.

The main features of each crash as identifi d by analyzing the crash reports were as follows: the location where the crashes happened, the number of crashes, injuries, and fatalities, type of crash, type, and number of vehicles involved, road surface conditions, lighting conditions, marking conditions, the number of legs and lanes, lane width, $AADT_{maj}$, that is, the AADT on major roads in terms of vehicles per day, and $AADT_{min}$, that is, the AADT on minor roads in terms of vehicles per day, the presence of left- urn lanes, median-refuge islands, right-turn lanes, and the diame- ter of the roundabouts.

A total of 827 crashes were recorded in 5 years. The geometric features of each investigated intersection (see **Table 1**) were established from documents made available by the Regional Administrative Offices A total of 770 crashes were observed at non-circular intersections, 623 of which were injury crashes, and 147 involved property damage only (PDO) crashes; a total of 1025 injuries were recorded at non-circular intersections, and 12 fatalities occurred. A total of 57 crashes were observed at single-lane roundabouts, of which 36 were injury crashes and 21 PDO crashes; a total of 57 injuries were recorded at single-lane roundabouts, and no fatalities occurred.

The crash value for each intersection shows the number of crashes over a 5-year study period, while the frequency of injury crashes refers to the number of injury crashes per year at each intersection during the study period. **Figure 2a** shows that

Features at intersection	Non-circular intersections				Roundabouts			
	Min	Mean	Max	C.V.	Min	Mean	Max	C.V.
Total number of crashes	3	4.51	10	0.84	1	1.32	5	0.75
Total number of injury crashes	0	0.99	6	0.79	0	0.86	2	0.44
Number of injuries	0	1.61	7	0.97	0	1.32	4	0.89
Frequency of crashes per year	0.33	0.48	4.33	0.81	0.12	0.18	0.82	0.77
Frequency of injury crashes per year	0	0.22	1.38	0.79	0	0.10	0.13	0.66
Frequency of injuries	0	0.28	2.00	0.93	0	0.21	0.5	0.87

Note: Min—minimum value; Mean—average value; Max—maximum value; C.V.—coefficient of variation, equal to the standard deviation divided by the mean value.

Table 1.
Overview of the main statistical features of the crashes and the intersection type.

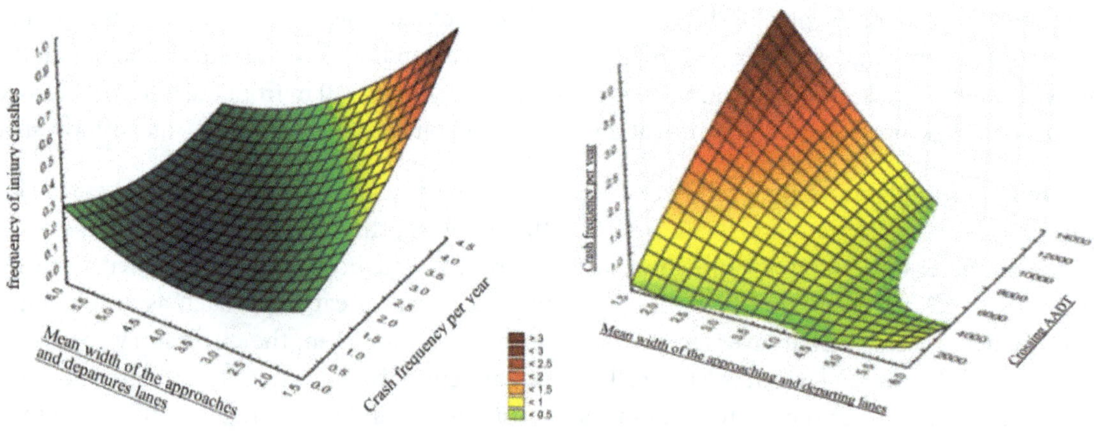

a) b)

Figure 2.
Hazard maps of injury crash frequency. (a) Crashes vs geometric properties of lanes. (b) Crashes vs traffic and geometric properties.

injury crash frequencies per year increase when the total crash frequency per year increases and the mean width of the approaching and departing lanes of the regular intersections decreases. **Figure 2b** shows how crash frequencies per year increase when the AADT crossing the intersection increases and when the mean width of the approaching lane to an intersection or departing lane from the intersection decreases.

Data analysis: evaluating measures reflecting crash risk exposure

Calculating crash rate at intersections as a first measure of safety level

In the light of the research goals set out in Section 1 and shown in **Figure 1**, a procedure developed in 1995 by the Italian National Research Council [12] was used to assess the safety level of traffi conditions at each ith intersection investigated, as shown in the fl wchart in **Figure 3.**

Before computing the crash safety level, LoS and total crash cost, a technique for filtering anomalous crash rates was adopted using the 3σ method. The method is based on the calculation of the standard deviation (σ) and mean values (μ) for crash rate distribution to check the homogeneity of scattering around the average and the

Assessing the crash rate for the i-th study intersection by adopting Equation 1 (see section 1.2.3 in National Research Council, 1995)

$$CR_i = \frac{10^6 \cdot N_i}{365 \cdot AADT_{total\ entering,i}} \qquad (1)$$

N_i is total number of crashes during the study period;
$AADT_{total\ entering,i}$ is the total Annual Average Daily Traffic entering at the intersection during the study period as whole, vehicles per day

⇩

calculating the mean value of the crash rate (C_{mean}) by involving all the intersections

⇩

calculating the threshold values "lower limit $CR_{lower\ limit}$ and upper limit $CR_{upper\ limit}$" of the crash rates that are the extremes of the 95% confidence interval by using a Poisson distribution for the crashes by adopting the Equations 2 and 3 (see section 1.3 in National Research Council, 1995)

$$CR_{lower\ limit} = CR_{mean} - K \cdot \sqrt{\frac{CR_{mean}}{365 \cdot AADT_{total\ entering,\ i}}} - \frac{1}{2 \cdot 365 \cdot AADT_{total\ entering,\ i}} \qquad (2)$$

$$CR_{upper\ limit} = CR_{mean} + K \cdot \sqrt{\frac{CR_{mean}}{365 \cdot AADT_{total\ entering,\ i}}} + \frac{1}{2 \cdot 365 \cdot AADT_{total\ entering,\ i}} \qquad (3)$$

K is the constant of the Poisson distribution probability equal to 1.645 associated with 10% of error probability

⇩

defining the crash level of each i-th intersection by adopting the Equations 4, 5 and 6 (see section 1.3 in National Research Council, 1995)

$$CR_i \le CR_{lower\ limit} \Rightarrow \text{low crash level} \qquad (4)$$

$$CR_{lower\ limit} < CR_i \le CR_{upper\ limit} \Rightarrow \text{medium crash level} \qquad (5)$$

$$CR_i > CR_{upper\ limit} \Rightarrow \text{high crash level} \qquad (6)$$

Figure 3.
Assessment of the crash safety level of each ith study intersection.

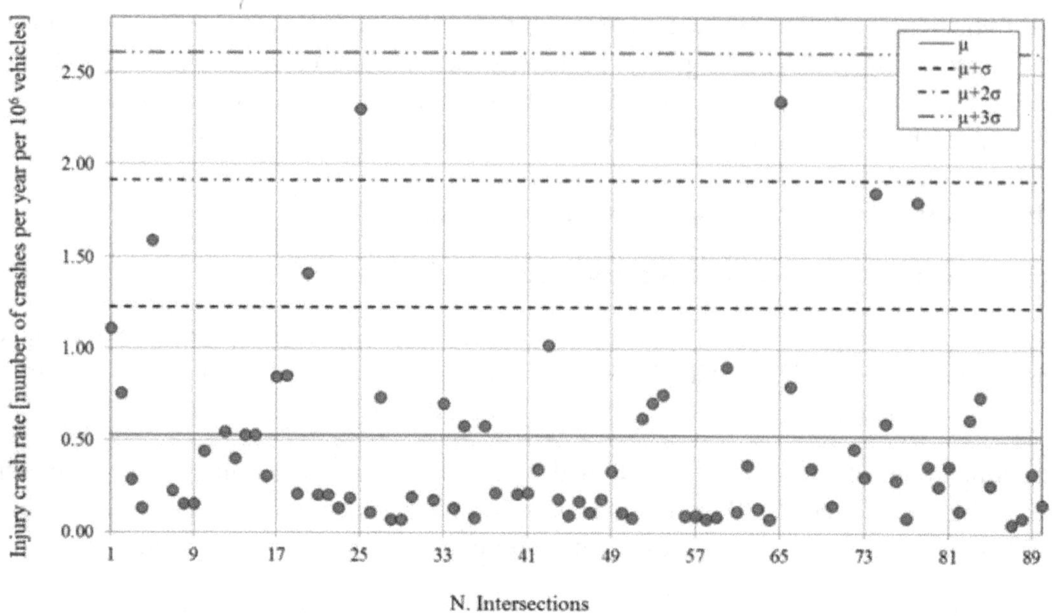

Figure 4.
Control charts of the crash rate values for typical intersections.

maximum deviation at 3σ. Figure 4 shows an example of the control chart of the crash rates for non-circular intersections throughout the study period. It can be observed how 92% of the measurements fall within the range [0; μ + σ] = [0; 1.24], 97% fall within the range [0; μ + 2σ] = [0; 1.94], and all the values fall within the range [0; μ + 3σ] = [0; 2.64]. $CR_{lower\ limit}$ is equal to 0.36 crashes per year per 10^6 vehicles crossing the ith intersection, and $CR_{upper\ limit}$ is equal to 0.52 crashes per year per 10^6 vehicles crossing the ith intersection.

The overall results show that 63% of the total number of intersections indicate a low crash level (a total of 51 non-circular intersections and 7 single-lane round-abouts); 33% show a high crash level (a total of 34 non-circular intersections), and the remaining 4% represents a medium crash level (four non-circular intersections).

Calculating level of service as a second measure of safety level

The next step in the study focused on assessing LoS and crash costs for the four intersections where a high crash level and medium-high crash cost were observed. Neither the non-circular intersections respecting the Italian Road Design Standard nor the roundabouts are included among the "black" rankings. **Figure 5** shows an excerpt of the current geometric design of four "black" ranking intersections.

Table 2 shows the main features of the investigated intersections mentioned above: the number of legs, $AADT_{maj}$, $AADT_{min}$, and CRi, as well as the total number of crashes, the total number of injuries, and the total number of vehicles damaged during the collision. **Table 3** shows the distribution of the hourly traffi fl w (q_j) for the equivalent passenger cars in the different travel directions as illustrated in **Figure 5**.

The geometrical configur tion of the four study intersections in **Figure 5** is very simple, and no additional geometric modules exist to promote safe maneuvering, according to the specifi ations in Section 2. Th s is the opposite of what happens at intersections where a "low crash level" was observed, where additional modules exist, and where geometric features respect the Italian Road Design Standard requirements in full. LoS was assessed for the entire intersection by evaluating control delay dj for each maneuver jth. The HCM2016 [13] defi es control delay as the measure of effectiveness used to set LoS at TWSC intersections as perceived by users.

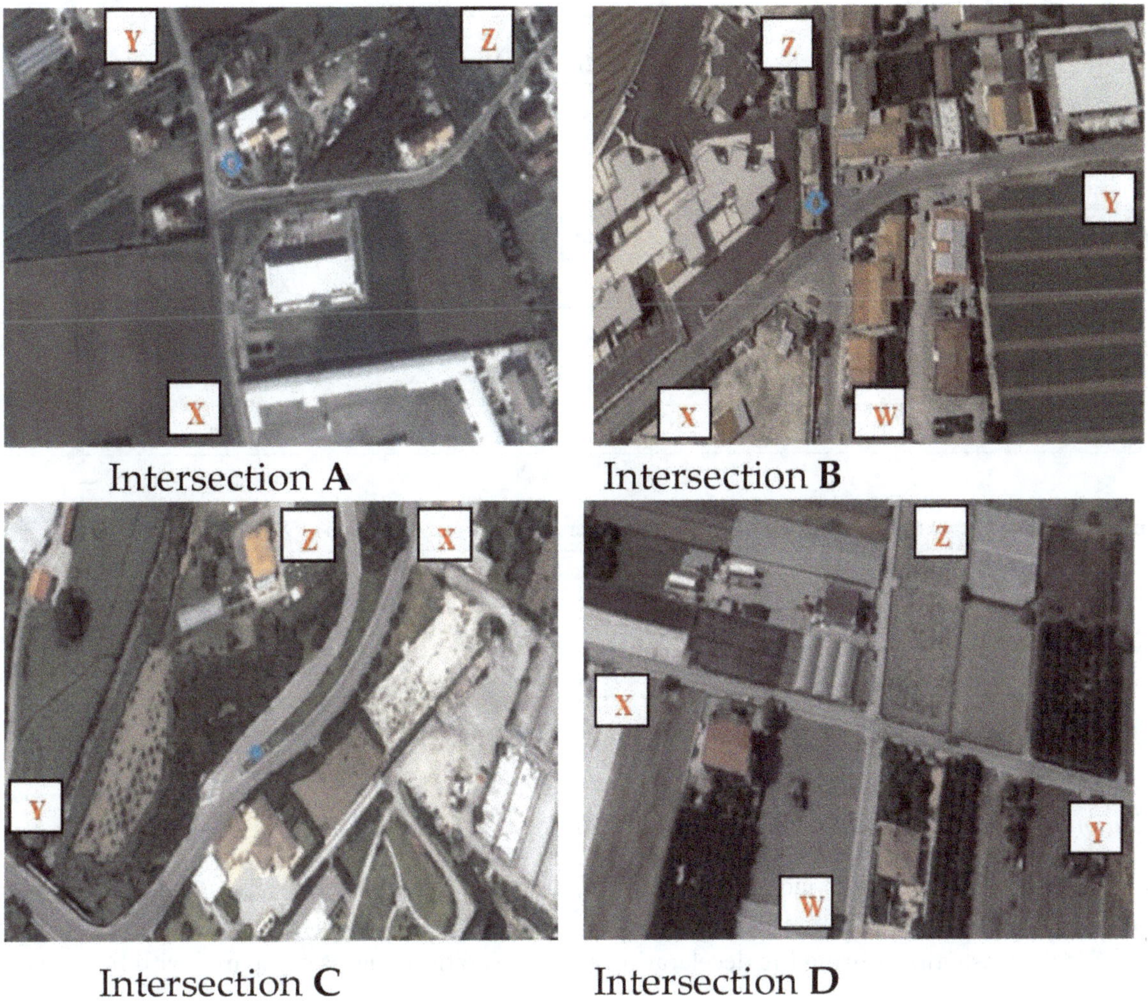

Figure 5.
The geometric design of the currently non-circular intersections studied.

Site	Number of legs	$AADT_{maj,}$	$AADT_{min,}$	CR_i	Number of Crashes	Injuries	Total number of Vehicles damaged during collision
		vpd	vpd		5-year study period		
Intersection A	3	3163	2120	2.22	4	3	4
Intersection B	4	3518	3153	1.32	11	4	9
Intersection C	3	6193	3097	1.72	12	8	7
Intersection D	4	4112	3102	2.09	14	4	6

Table 2.
Overview of the crash and traffic features of four typical intersections investigated.

The analytical model (Eq. (7)) adopted to estimate average control delay dj per maneuver jth refers to Eqs. (17)–(38) of HCM2016 [13] and assumes that there is no residual queue at the start of the analysis period. In most cases, the recommended analysis period is 15 min.

$$d_j = \frac{3600}{C_{e,j}} + 900 \cdot T \left[\frac{q_j}{C_{e,j}} - 1 + \sqrt{\left(\frac{q_j}{C_{e,j}} - 1 \right)^2 + \frac{\frac{3600}{C_{e,j}} \cdot \frac{q_j}{C_{e,j}}}{450T}} \right] + 5 \tag{1}$$

Site	Traffic volume, vph				Site	Traffic volume, vph				
Intersection A	Direction	X	Y	Z	Intersection B	Direction	X	Y	Z	W
	X	—	226	13		X	—	150	60	60
	Y	170	—	68		Y	100	—	70	100
	Z	70	11	—		Z	70	90	—	80
						W	100	30	110	—
Intersection C	Direction	X	Y	Z	Intersection D	Direction	X	Y	Z	W
	X	—	374	42		X	—	150	60	80
	Y	274	—	190		Y	180	—	70	80
	Z	142	90	—		Z	70	40	—	60
						W	30	50	80	—

Table 3.
Distribution of the hourly traffic volume in the subject lane per maneuver.

where q_j is the fl w in the subject lane for maneuver j, in vph; $C_{e,j}$ is the effective capacity of the subject lane for maneuver j, in vph; T is the time period, in hours (T = 1 for a 1-h analysis, T = 0.25 for a 15 min analysis); and 5 is the waste time during the deceleration and acceleration phases compared with free fl w speed, expressed in seconds (5 s).

Eq. (7) was adjusted for the real context working on one of the main variables: effective capacity $C_{e,j}$.

$C_{e,j}$ was calculated by adopting real measurements of $t_{c,j}$ (critical gap per maneuver j-th) and $t_{f,j}$ (follow-up time per maneuver j-th) values at the four study intersections shown in **Figure 5** instead of adopting the HCM2016 equation based on studies across the United States.

The evaluation of $t_{c,j}$ critical gaps is not immediate and can appear difficult to estimate from a fi ld measurements sample; results of empirical studies have shown that different combinations of configur tion and situational influences may lead to diverse profiles of compliance and proactive safety behavior among drivers [21].

In the literature, several techniques exist to calculate gap acceptance data assuming the consistency of road drivers, for example, the Raff nd Hart method [22], the Drew-Dawson method [23–25], and the stepped line model.

As a result, the $t_{c,j}$ value for each driver entering an intersection from a minor road in **Figure 5** was calculated using the Drew-Dawson method, based on the median time value. **Figure 6** shows an example of critical gaps ($t_{c,j}$) for drivers crossing intersection A (see **Figure 5**) turning from leg Y to leg Z (left turn from minor to major road) and from leg Z to leg Y (right turn from minor to major road) during time period T. **Table 4** shows an overall view of the observed values of the $t_{c,j}$ and $t_{f,j}$ variables for the maneuvers at intersection A.

The control delay for an entire intersection $d_{entire\_intersection}$ (see Eq. (8)) is calculated by computing a weighted average of the control delay for each maneuver dj, weighted by the volume of each fl w for the maneuver investigated.

$$dentire\ intersection = \frac{\sum d_j q_j}{\sum q_j} \tag{2}$$

According to the thresholds defi ed in HCM2016 [13], the LoS was defi ed for each maneuver by also associating qualitative measures from A (control delay between 0 and 10 s/veh) to F (control delay more than 50 s/veh) as provided in

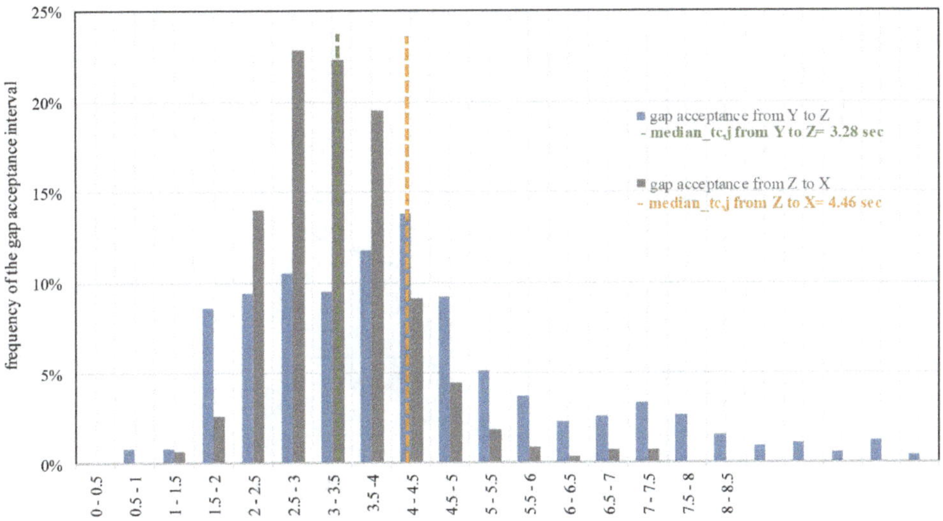

Figure 6.
Example of $t_{c,j}$ assessment using the Drew-Dawson method for two maneuvers at intersection A.

$t_{c,j}$	X	Y	Z	$t_{f,j}$	X	Y	Z
X	—	<1 s	<1 s	X	—	<1 s	<1 s
Y	<1 s	—	3.28 s	Y	<1 s	—	1.76 s
Z	6.62 s	4.46 s	—	Z	3.87 s	2.53 s	—

Table 4.
Observed $t_{c,j}$ and $t_{f,j}$ values for all maneuvers at intersection A.

Study intersection	Control delay ($d_{entire\ intersection}$)	Level of Service (LoS)	Total crash cost (TTC)
	s/veh		EUR
A	14	B—vehicle control delay 10–15 s	251,637
B	26	D—vehicle control delay 25–35 s	363,698
C	33	D—vehicle control delay 25–35 s	642,850
D	17	C—vehicle control delay 15–25 s	340,640

Table 5.
Overview of control delay and crash costs at non-circular intersections with the old configuration.

Exhibit 17-2 of the HCM2016 [13]. **Table 5** shows the dj values for the intersections investigated as listed in **Tables 2** and **3** and the corresponding crash costs from the cost of an injured person approximately equals to 73,631 Euro, fatality equals to 1,394,434 Euro, and damaged vehicles almost of 7686 Euro.

Table 5 shows that the total crash cost (TCC) increases when the control delay of the entire intersection $d_{entire\ intersection}$ investigated increases: when the mean dj of the vehicles leaving the intersection area increases, this indicates that drivers do not feel safe to make the maneuver. Th s circumstance is mainly due to the poor geometric configur tion of the intersection and, as confi med by the preliminary results of this study, can amplify the frequency and severity of crashes.

Figure 7 shows a positive linear relationship using the ordinary least square method to predict the total crash cost in euros by varying the average control delay at the entire intersection, which is a measure of LoS (Eq. (9)). The parameters

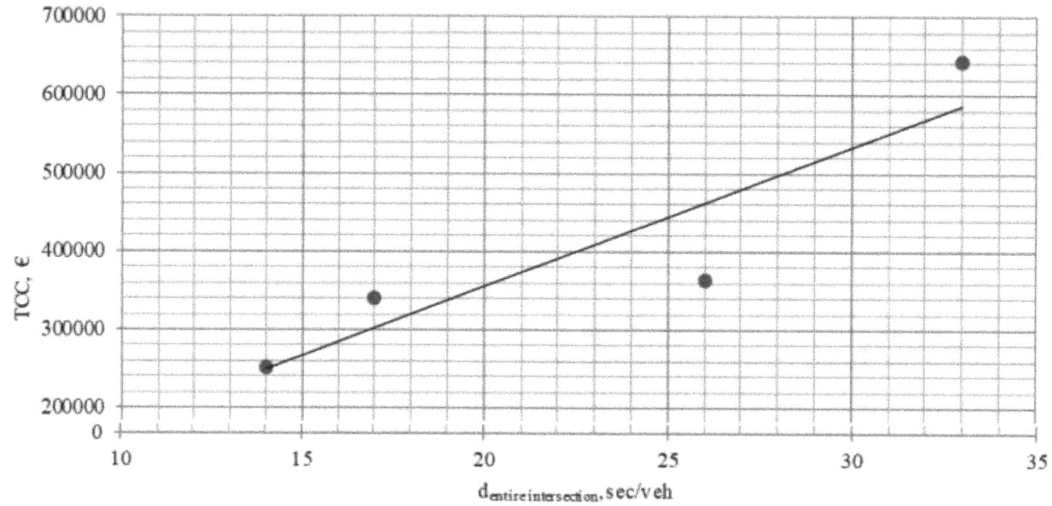

Figure 7.
Total crash cost versus average control delay of the entire intersection.

Std. error	t-value	p-value	Lower confidence limit	Upper confidence limit
1457.90	12.19	0.001189	13130.97	22410.36

Table 6.
Statistical parameters of the regression model.

included in the TCC prediction model are signifi ant with a 95% level of confide ce (**Table 6**). The adjusted coeffici t of determination (R^2) of the model is 83%.

$$TCC = 17:771\ d_{entire\ intersection} \tag{3}$$

Treatment to manage and reduce the crash level by carrying out strategies able to improve the expected LoS and safety levels

The next step in the research was to design different geometric solutions (**Figure 8**) for intersection D (**Figure 5**), which has a high crash rate, medium-high crash cost, low-medium LoS (**Table 5**) and to estimate, on the one hand, the expected LoS and, conversely, the expected reduction of the annual crash frequency of the configur tions hypothesized. The solution shown in **Figure 8a** is an adaptation of intersection D to the requirements of the Italian Road Design Standard on the geometric design of road intersections [14]. The solution shown in **Figure 8b** refers to transforming the shape into a compact roundabout.

Control delay: comparing before and after solutions

Before redesigning the geometric configur tion of the area of intersection to include a roundabout, for which the investigated database showed that the mean crash rate over 5 years was lower than at non-circular intersections, it was decided to adjust the current non-circular intersection in line with the requirements of the Italian Standard in force in two phases:

- **Phase I:** adjusting the radius of the edges of the entry and exit legs of the intersection, the width of the traffi lanes, the removal of obstacles in the areas within the so-called sight triangles, and the addition of median-refuge islands on the major and minor roads, as recommended by the Italian Road Design Standard [14];

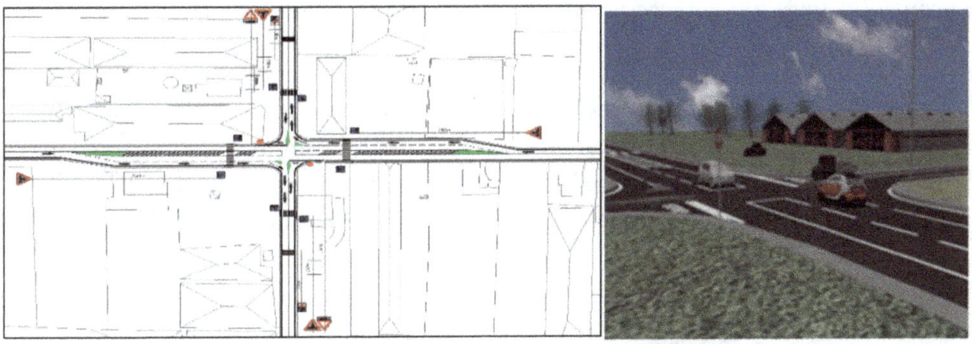

a)

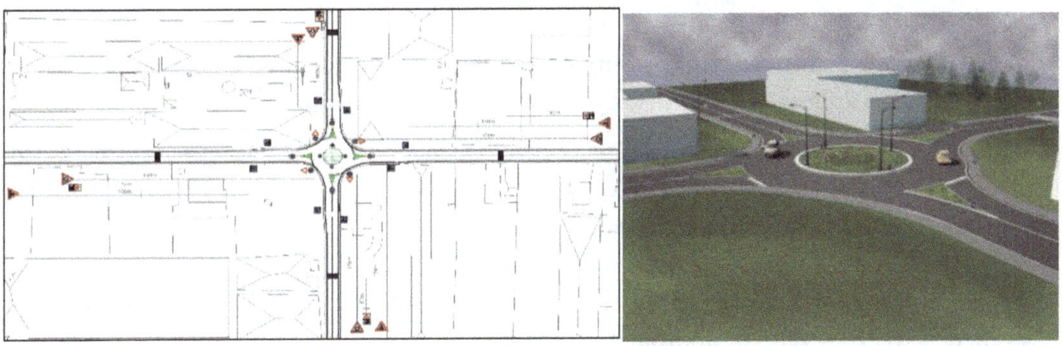

b)

Figure 8.
Advanced geometric design solutions for an existing non-circular intersection. (a) Adjustment according to the Italian Standard. (b) Changing the configuration into a compact roundabout (Dext = 26 m) according to the Italian Standard.

- **Phase II:** adding left- urn lanes on major-road approaches based on the ratio between the volume of vehicles turning left er hour and the total traffic olume on the highway per hour.

Right-turn lanes on minor road stops are not permitted by the Italian Design Standard [14], and they are not included in the design.

The intersections investigated are almost totally equipped with right-turn lanes on major roads and they have lighting within and approaching the intersection area; consequently new lighting and new right-turn lanes on major roads are not required. Control delay at the entire intersection for the fi st advanced geometric solution of intersection D (**Figure 8a**) was estimated as in Section 3.2; the expected LoS of this geometric adjustment is shown in **Table 7**. In particular, to compute the

Non-treatment site		**Expected configurations**		
Control delay on the entire intersection, s/veh	**LoS before treatment**	**Site**	**Control delay on the entire site, s/veh**	**LoS after treatment**
17	C	Intersection*	9	A
		Compact roundabout**	4	A

Adjusted to the Italian Standard without changing the shape by adding further modules* ***Figure 8a.*
***Figure 8b.*

Table 7.
*Comparison of the control delays at advanced geometric solutions with those of intersection D (**Figure 4**, old configuration).*

average control delay for the roundabout approach as a whole in order to make comparisons with other intersection types, control delay d_j for the ith approach was calculated by computing a weighted average of the delay for each lane on the approach (Eq. (7)), weighted by the volume in each lane. The calculation is shown in Eq. (10) using SETRA diagrams for the expected control delay at each maneuver, as the calibration conditions refl ct the actual study context.

$$d_{\text{approach, i}} = \frac{d_{\text{left lane}} \times q_{\text{left lane}} + d_{\text{right lane}} \times q_{\text{right lane}}}{q_{\text{left lane}} + q_{\text{right lane}}} \tag{4}$$

The control delay $d_{entire\ roundabout}$ for the entire roundabout is similarly calculated by computing a weighted average for the delay at each approach, weighted by the volume on each approach and represented by Eq. (11):

$$d_{entire\ roundabout} = \frac{\sum d_{approach,\ i}\, q_i}{\sum q_i} \tag{5}$$

where $d_{entire\ roundabout}$ is the control delay for the entire roundabout, s/veh; $d_{approach,k}$ is the control delay for approach k^{th}, s/veh; q_i is the fl w rate for approach i^{th}, vph.

Control delay at the entire intersection: comparing before and after solutions

It is imperative for a designer to understand the relationships between design features and crash frequency.

The effectiveness of all the changes that have been designed without changing the shape of a regular intersection but adjusting it to the Italian Standard (see **Figure 8a** and Section 4.3.1) was confi med by the expected crash frequency values computed by adopting the SPF available in Biancardo et al. [16]. Biancardo et al. [16] worked in line with HSM2010 [17] procedure and revised the equation available in the Manual to predict crash frequency at three and four-leg rural unsignalized at-grade intersections.

Th N_{spf} formulation [16] was here used (see Eq. (12)) to predict the crash frequency at the two-lane two-way four-leg intersections studied in greater depth (intersection D) as it was calibrated using a data set that adequately refl cts and partly overlaps with what is explored here. MLW is the mean lean width of the approaching and departing lanes.

$$N_{spf} = AADT \exp\left[-1.042\, MLW - 8.5\right] \tag{6}$$

Eq. (12) applies to an $AADT_{maj}$ range from 0 to 14,700 vpd and $AADT_{min}$ range from 0 to 3500 vpd.

In HSM2010 [17], Crash Modifi ation Factors (CMFs) are introduced to account for the specific site conditions that differ from the hypothesized base conditions. Under base conditions, the CMF is 1.00 (i.e., **Figure 5d**), while the CMF is less than when a geometric configur tion in compliance with the Standard and with many additional modules exists and, consequently, a reduction of average yearly crash frequencies can be expected. $N_{predicted}$ (predicted average crash frequency for a specific year for site type x) is shown in Eq. (13), where the effect of the skew angle does not appear, as study intersection D has an 80° angle, very close to orthogonal road axes and, consequently, no additional benefits can derive from further correction or the right-turn lanes that already exist on major roads.

$$N_{predicted} = N_{spf}\, CMF_{LTL} \tag{7}$$

where N_{spf} was determined by the following Eq. (12).

CMF_{LTL} was computed using the HSM procedure, and it benefits from the effects of the presence of left- urn lanes (LTL) on the major road, specifi ally in terms of expected average annual crash frequency reduction compared with what can be observed at intersections with a poor geometric configur tion. CMF_{LTL} is equal to 1 for four-leg unsignalized rural intersections that meet base conditions. It equals 0.13 for the left- urn lanes present [16].

Table 8 shows, in the light of the foregoing, the expected annual number of crashes if the intersection is adjusted to Italian Road Design Standard [14] requirements by introducing additional geometric modules as listed in the fi st part of Section 3.3.1.

Moving on now to the evaluation of the effectiveness of the second treatment (the conversion of typical intersections into compact roundabouts) suggested for intersection D in order to check whether the level of exposure to crash risk can be reduced and is generally well managed, the EB procedure was adopted, as mentioned in the Literature review section.

First of all, it is necessary to calculate the expected annual number of crashes (m) if conversion to a roundabout does not take place. Eq. (14) was adopted to obtain a site-specific estimate of the m variable at a typical intersection before conversion to a roundabout:

$$m = w_1 x þ w_2 P \tag{8}$$

where m is the expected site-specific annual number of crashes or injury crashes before conversion; x is the count of crashes in the n years before conversion (see **Table 2**, a total of 14 crashes occurred in 5 years with 4 injuries); n = 5 is the study period in this research; w^1 and w^2 are weights, Eqs. (15) and (16) [18]:

$$w_1 = \frac{P}{\frac{1}{k} + nP}$$

$$w_2 = \frac{\frac{1}{K}}{\frac{1}{k} + nP}$$

where P is the prediction of the annual number of crashes, or the annual number of injury crashes depending on what it is necessary to investigate using an SPF to identify intersections with similar characteristics before conversion; k is the dispersion parameter for a given model, estimated from the SPF calibration process using a maximum likelihood procedure.

Case study	Crashes per year	
Intersection D	Total	Injury
Expected annual number of crashes	1.99	0.57
Expected annual changes to the number of crashes	-0.81	-0.23
Reduction in crashes	-28%	-29%

Table 8.
Calculation of the expected change to the number of crashes after shape adjustment in line with the requirements of the Italian Road Design Standard.

Case study—Intersection D		Crashes per year
	Total	Injury
BEFORE (neither adjustment nor conversion) (Eq. (14))		
P	0.89	0.20
K	0.77	1.25
w_1	0.15	0.11
w_2	0.23	0.45
m	2.37	0.64
AFTER (after conversion to roundabout—solution 2)		
Expected annual number of crashes	1.86	0.26
Expected changes in no. of crashes	-0.51	-0.38
Reduction in no. of crashes	-21%	-59%

Table 9.
Calculation of the expected change in the number of crashes when intersection D is converted into a roundabout.

Rodegerdts et al. [18] suggested k equals 0.77 for an SPF that predicts the total number of crashes per year, and k equals 1.25 for an SPF that predicts the total number of injury crashes per year. In chapter C, [18] are defi ed the results of the efforts to develop intersection and approach-level models. These models relate crash prediction to the number of lanes, number of legs, and the average annual daily traffic SPFs used to predict the expected total number of crashes per year at intersection (Eq. (17)) or the expected total crash injuries per year at intersection (Eq. (18)) are as follows:

$$P = \exp(-8{:}63)\,(AADT_{\text{total entering}})^{0.952} \tag{11}$$

$$P = \exp(-8{:}733)\,(AADT_{\text{total entering}})^{0.795} \tag{12}$$

Eqs. (17) and (18) have been used in this study to predict m variable, since they were validated using the data set that is adopted here as shown in [16].

Eq. (19) was adopted to predict the expected total crash frequency per year after converting the intersection into a single-lane roundabout [19], where $AADT_{total\ entering}$ is the total annual average daily traffi entering the roundabout, equal to 7642 vpd for the intersection in question.

$$m = 0{:}023\ (AADT_{total\ entering})^{0:749} \tag{13}$$

The expected safety effects are shown in **Table 9.**

Results and discussions

A comparison of the expected crash frequency between conversion and non-conversion into a single-lane roundabout of the four-leg two-way-stop intersection is performed by plotting **Figure 9**. This makes it possible to identify a maximum threshold for the $AADT_{total\ entering}$ at the single-lane roundabout when this configuration replaces an existing typical intersection without damaging the required safety levels.

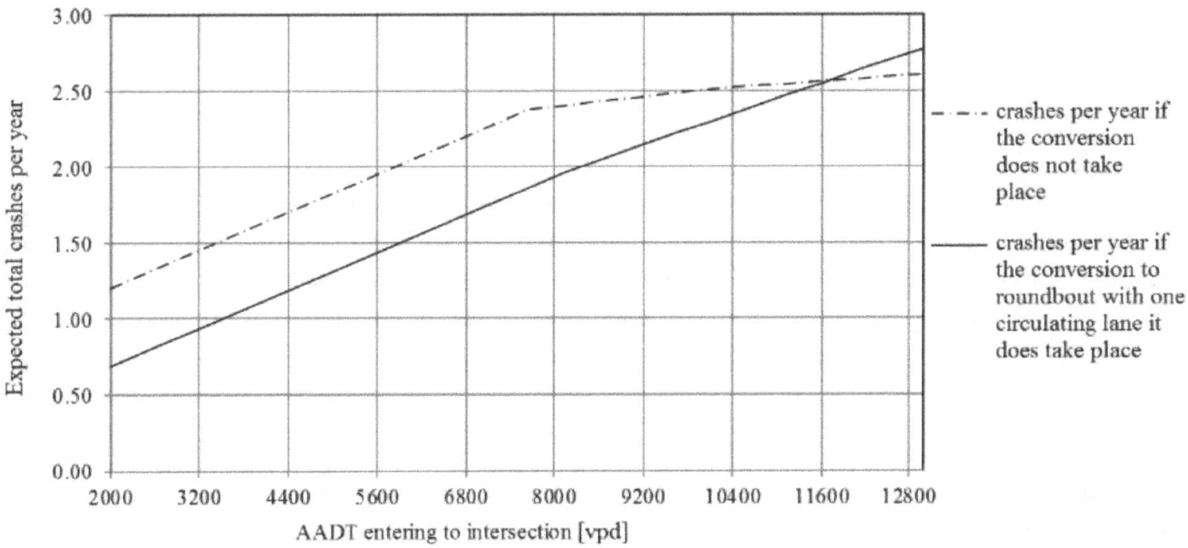

Figure 9.
Roundabout performance in terms of expected crash frequency for $AADT_{total\_entering}$.

The results summarized in **Table 10** and listed below highlight a strong correlation between the LoS and the safety level for managing hotspots along road net- works and the corresponding crash risk levels, improving system quality for users. The results achieved show that, by increasing control delay throughout the entire intersection, the expected safety level for the expected annual number of crashes decreases. Conversely, when the estimated level of service increases (reducing the control delay of the entire intersection), the safety level improves (translating into a low value for the expected value of annual crashes per year). Results confirm that if a study intersection, under specific traffic conditions, and in a specific environmental and surrounding context, has a suitable and correct geometric configuration for reducing the number of conflict points during possible maneuvers, the control delay on the entire structure is reduced, and the LoS improves. This reflects indirectly, but positively, on the safety level because the expected value of the annual crashes decreases.

Th s research aimed to identify road strategies to improve road safety conditions at rural two-lane two-way intersections with stop-control in order to identify crash risk factors that may affect the Level of Service (LoS) and the safety level of the

Study case Intersection D	**Before**	**After**	
	No treatment site	**Solution No. 1**	**Solution No. 2**
		Conversion to compact roundabout	**Adjustment to the Italian Standard by designing additional geometric modules**
Expected control delay of the entire intersection	17 s/veh	4 s/veh	9 s/veh
Expected annual number of site-specific cr shes	2.37	1.86	1.99
Level of Service (LoS)	C	A	A
Reduction in total crashes per year		21%	16%
Reduction in injury crashes per year		59%	31%
Annual economic benefit in crash savings		29,000 EUR	16,000 EUR

Table 10.
LoS and expected crash frequency at advanced geometric solutions.

road system on the one hand, and to analyze the effectiveness of treatment for the effective management of hotspots and ensure the good operation of the system, on the other hand. The procedure investigated can help in the allocation of resources according to the needs and severity of a possible crash event that, although rare, can have dramatic consequences, especially when risk factors are not identifi d, analyzed, and reduced.

Conclusions

In this chapter, a methodological process that can also be implemented in other domains was shown to calculate, manage, and reduce, through appropriate treatments, the expected crash risk level measured in terms of yearly crash frequency and Level of Service.

First of all, the procedure aimed to identify, and then manage, the hotspots on a rural road intersection network where high exposure to crash risk can be observed. It also sought to rank the hazardous sites, for which two measures of exposure to risk were suggested and assessed in line with the research presented, namely the crash rate and Level of Service in terms of control delay at the intersection area.

Of course, a safe system approach requires a fundamental cultural and ethical shift in thinking, but it is also true that the current road transport system is not as safe as it could be. However (a) if the system could be well supervised, (b) if the trend of a number of system status indicators (i.e., crash rate level, level of service, crash cost, etc.) could be carefully plotted to check their decay over time, (c) if design errors were promptly identifi d, and (d) if the correlations between design errors/access management and factors that cause increased exposure to crash risk were then investigated, in the event of human error or driver distraction, the resulting severity might not be as high. Obviously, system designers and system users must all share responsibility for managing crash forces to a level that does not result in death or serious injury.

It has been verifi d whether improvements can be achieved in terms of safety level (reduction of the number of crashes and injuries) and the quality of traffi (reduced control delay over the entire intersection) when the geometric design of existing intersections belonging to two-lane rural roads and located on a flat area does not meet the Italian Standard.

The experimental method covered two parallel trajectories that ultimately converge:

- adapting an existing at-grade intersection without changing its shape;
- changing its geometry according to the Italian standards, keeping traffi features and environmental conditions constant.

The results show that for the intersections in question, designing a single-lane roundabout according to the Italian Road Design Standard, or an intersection introducing left- urn lanes, deceleration lanes, and median-refuge islands could help to achieve this goal. Compact roundabouts are, in any case, the best solution in terms of Level of Service and safety level because they contribute to strongly reducing delay as well as crashes.

Conflict of interest

No potential confli t of interest was reported by the authors.

Author details

Francesca Russo*, Salvatore Antonio Biancardo and Rosa Veropalumbo
University of Naples Federico II (NA), Italy

*Address all correspondence to: francesca.russo2@unina.it

References

[1] Dickey BD, Santos JR. Risk analysis of Safety Service Patrol (SSP) systems in Virginia. Risk Analysis. 2011;31(12): 1859-1871

[2] Freiria S, Tavares AO, Juliao RP. The multiscale importance of road segments in a network disruption scenario: A riskbased approach. Risk Analysis. 2014; 35(3):484-500

[3] Directive 2008/96/EC of the European Parliament and of the Council of 19 November 2008 on Road Infrastructure Safety Management OJ L 319. November 29, 2008. pp. 59-67. Available from: https://eur-lex.europa.e u/legal-content/EN/TXT/PDF/?uri= CELEX:32008L0096&from=EN

[4] Persaud B, Retting RA, Garder P, Lord D. Safety effects of roundabout conversions in the United States: Empirical Bayes observational before-after study. Transportation Research Record. 2001;1751:1-8

[5] Elvik R. Effects on road safety of converting intersections to roundabouts: Review of evidence from non-U.S. studies. Transportation Research Record. 2003;1847:1-10

[6] Hauer E. The Art of Regression Modelling in Road Safety. New York: Springer; 2015

[7] Kruschke JK. Bayesian assessment of null values via parameter estimation and model comparison. Perspectives on Psychological Science. 2011;6(3): 299-312

[8] Congdon P. Bayesian Statistical Modelling. Wiley Series in Probability and Statistics. Chichester: Wiley; 2001

[9] Lee PM. Bayesian Statistics: An Introduction. London: Arnold; 1997

[10] Asendorpf JB, Conner M, De Fruyt F, De Houwer J, Denissen JJA, Fiedler K, et al. Recommendations for increasing replicability in psychology. European Journal of Personality. 2013;27:108-119

[11] Xie K, Ozbay K, Kurkcu A, Yang H. Analysis of traffi crashes involving pedestrians using big data: Investigation of contributing factors and identifi ation of hotspots. Risk Analysis. 2017;37(8):1459-1476

[12] Consiglio Nazionale delle Richerche. Criteri per la Classifi azione della Rete delle Strate Esistenti ai Sensi Dell'art. 13, Comma 4 e 5 del Nuovo Codice della Strada. Allegato 2. Metodologia per la Verifi a delle Condizioni di Sicurezza ai Fini della Classifi azione in Deroga delle Strade Esistenti. 1998. Available from: h ttps://www.docenti.unina.it/webdoce ntibe/allegati/materiale-didattico/ 67022

[13] Transportation Research Board. Highway Capacity Manual, 6th Edition: A Guide for Multimodal Mobility Analysis. Washington, DC: The National Academies of Sciences, Engineering, and Medicine; 2016

[14] Ministero delle Infrastrutture e dei Trasporti. Norme Funzionali e Geometriche per la Costruzione delle Intersezioni Stradali - GU n. 70 del 24-7- 2006. 2006. Available from: http:// www.mit.gov.it/mit/mop_all.php?p_id= 13799

[15] Candappa N, Logan D, Van Nes N, Corben B. An exploration of alternative intersection designs in the context of safe system. Accident; Analysis and Prevention. 2015;74:314-323

[16] Biancardo SA, Russo F, Zilioniene D, Zhang W. Rural two-lane two-way three-leg and four-leg stop-controlled intersections: Predicting road safety effects. The Baltic Journal of Road and Bridge Engineering. 2017;12(2):117-126

[17] American Association of State Highway and Transportation Offi als (AASHTO). Highway Safety Manual. Washington, DC: AASHTO; 2010

[18] Rodegerdts L, Blogg M, Wemple E, Myers E, Kyte M, Dixon M, et al. Roundabouts in the United States, National Cooperative Highway Research Program Report 572. Washington D.C.: Transportation Research Board of the National Academies; 2007. 125p. Available from: http://nacto.org/docs/ usdg/nchrprpt572.pdf

[19] Rodegerdts L, Bansen J, Tiesler C, Knudsen J, Myers E, Johnson M, et al. Roundabouts: An Informational Guide, Second Edition, National Cooperative Highway Research Program Report 672. Washington D.C.: Transportation Research Board of the National Academies; 2010. 396p. Available from: http://nacto.org/docs/usdg/nch rprpt672.pdf

[20] Ministero delle Infrastrutture e dei Trasporti. Norme Funzinali e Geometriche per la Costruzione delle Strade. D.M. n. 6792 05/11/2001. 2001. Available from: http://www.mit.gov.it/ sites/default/files/media/normativa/ 2016-02/Decreto_Ministeriale_numero_ 6792_05-11-2001_all_1.pdf

[21] Fugas CS, Silva SA, Melia JL. Profiling safety behaviors: Exploration of the sociocognitive variables that best discriminate between different behavioral patterns. Risk Analysis. 2013; 33(5):838-850

[22] Raff MS, Hart JW. A Volume Warrant for Urban Stop Signs. Saugatuck, Conn: Eno Foundation for Highway Traffi Control; 1950

[23] Drew DR. Traffi Flow Theory & Control. New York, United States: McGraw & Hill Series in Transportation. 1968

[24] Dawson RF. The hyperlang probability distribution—A generalized traffi headway model. Proceedings of the Fourth ISTTT in Karsruhe. Strassenbau und Strassenverkehehrstechnik. 1969;89: 30-36

[25] Mishra S, Khasnabis S. Optimization model for allocating resources for highway safety improvement at urban intersections. Journal of Transportation Engineering. 2012;138(5):535-547

5

A Critical Review on Population Synthesis for Activity-and Agent-Based Transportation Models

Ossama E. Ramadan and Virginia P. Sisiopiku

Abstract

Traditional four-step transportation planning models fail to capture novel transportation modes such as car/ridesharing. Hence, agent-based models are replacing those traditional models for their scalability, robustness, and capability of simulating nontraditional transportation modes. A crucial step in developing agent-based models is the defin tion of agents, e.g., household and persons. While model developers wish to capture typical workday travel patterns of the entire study population of travelers, such detailed data are unavailable due to privacy concerns and technical and fi ancial feasibility issues. Hence, modelers opt for population syntheses based on travel diary surveys, land use data, and census data. The most prominent techniques are iterative proportional fitting (IPF), iterative proportional updating (IPU), combinatorial optimization, Markov-based and fitness-based syn- theses, and other emerging approaches. Yet, at present, there is no clear guideline on using any of the available techniques. To bridge this gap, this chapter presents a comprehensive synthesis of practice and documents available successful studies.

Keywords: transportation planning, traffic s ulation, agent-based models, population synthesis

Introduction

Transportation simulation models are widely used for travel demand forecasting, testing design alternatives, or predicting travel behavior. In 1992, Axhausen and Gärling [1] developed a comprehensive review of conceptualizations and approaches of activity-based transportation models with special regard to the validity of behavioral assumptions of modeled population. In the course of their review, they concluded that individual travelers and households, rather than aggregates, should be identifi d and considered. Nevertheless, detailed travel records for individuals have never been easily accessible for several reasons, the most important being privacy issues and cost. Hence, individual travel diaries needed to be synthesized from travel surveys, census data, and publically available records. That process has since been known as population synthesis.

Population synthesizers initially were used as feeder data avenues to travel demand models [2]; however, recent shifts toward activity- and agent-based models brought population synthesizers to the spotlight, as they became determinants to the success or failure of any transportation model of that kind. Fitting is the core of any population synthesizer, with the main focus on fitting disaggregate sample of agents (represented by tabulated demographics of a representative sample of household and individual data) to aggregate constraints (represented by available aggregate data, such as data available from census). There are several approaches for fitting including iterative proportional fitting (IPF), iterative proportional updating (IPU), combinatorial optimization, Markov-based and fitness-based syntheses (FBS), and other emerging approaches [3]. The following sections present a critical review of each approach in the chronological order by which they were introduced to illustrate the progression and evolution of each approach, with emphasis on notable and well-established efforts.

Iterative proportional fitting approach

Iterative proportional fitting has been fi st introduced in 1940 by Deming and Stephan [4]. Since then, it became the foundation of population synthesis for transportation models and sometimes referred to as the Fratar technique [5]. The most notable realization of the IPF technique is attributed to Beckman et al. [6] who pioneered population synthesis efforts through their development of a methodology for creating a synthetic baseline population of individuals and households for microscopic activity-based models. Their technique relied on using census data represented by a Census Standard Tape File and Public Use Microdata Sample (PUMS) for a given Public Use Microdata Area (PUMA) of 100,000 individuals with matching variables. In their case, the marginal totals of a multiway table were known, and a sample from the population which generated those totals was provided; thus, they applied the IPF technique to develop constrained maximum entropy estimates of the true proportions in the population multiway table. Their rationale was built upon the consensus that IPF estimates maintain the same odds ratios as those in the sample table in the absence of any marginal information which was their case. To validate the population synthesis method, they compared demographic characteristics of the synthetic population with those of the true population using variables not involved in the population synthesis. Despite their pioneer effort, Beckman

et al. [6] did not provide an answer to the zero-cell problem in the PUMS; instead, they replaced it by 0.01 and imputed the corresponding household size. Müller and Axhausen [3] illustrated this as computing a series of tabulations $nij^{(k)}$, starting with the seed at $k = 0$, thus $nij^{(0)} = nij$ for all i rows and j columns. Furthermore, they illustrated how that series can be computed as represented by Eq. (1):

$$n_{ij}^{(k+1)} := n_{ij}^{(k)} \cdot \begin{cases} r_i \div n_{i\cdot}^{(k)} \\ c_j \div n_{\cdot j}^{(k)} \end{cases} \tag{15}$$

where $n_{i\cdot}$, is the row sum; $n_{\cdot j}$, is the column total; r_i, is the control total for row i; c_j, is the control total for column j.

Almost a decade later, Arentze et al. [7] addressed one of the limitations of the IPF methods, that is, generating synthetic households when the demographic data describes population in terms of individual counts. Their solution relied on devel- oping a two-step IPF procedure where, fi st, known marginal distributions of individuals are converted to marginal distributions of households of similar attributes and, second, the resulting marginal household distributions are used as constraints of a multiway household counts. Additionally, their approach aimed to assess the relevance of spatial heterogeneity across populations. The Dutch Albatross model was used as a case study and proof of concept. The validation results yielded sample biases in the synthetic population on the dimensions of socioeconomic class, the presence of children, and the availability of transport modes. However, they were able to resolve biases in overor underrepresentation of groups that were related to age and work status by fitting the relevant tables on these dimensions.

Simultaneous to the efforts of Arentze et al. [7], Guo and Bhat [8] addressed the two main drawbacks of IPF approach, namely, the zero-cell problem and the inability to control for statistical distributions of both household-and individual-level attributes. Additionally, their study aimed to enhance the scalability and generality of the IPF method as it required code-level changes that are cumbersome and skills that are not typically found within planning agencies, who are the typical users of such approach. The algorithm developed by Guo and Bhat [8] featured generic data structures and accompanying functions to avoid the zero-cell problem, as well as revisions to the algorithm of Beckman et al. [6] to allow simultaneous control of both household- and individual-level attributes. That generic algorithm was built upon an object-oriented architecture and contained eight major steps and a recurring procedure for merging any two contingency tables with common variables. The proposed approach was used to generate synthetic population for the Dallas-Fort Worth metropolitan area in Texas, and the statistical comparison yielded results that were closer to true population than that of Beckman et al. [6]. In addition, Guo and Bhat [8] concluded that a higher percentage deviation from target size (PDTS) yielded better balance at satisfying the household- and individual-level multiway distributions than lower values of PDTS.

Srinivasan et al. [9] went a step further and attempted to fi e-tune existing efforts to accommodate the household- and individual-level controls as well as assess the signifi ance of controlling individual-level attributes. That study was performed in support of Florida Department of Transportation (FDOT) efforts to incorporate sociodemographic attributes within the Florida Standard Urban Transportation Model Structure (FSUTMS). The research was motivated by the need for reduced aggregation errors, ensuring sensitivity to demographic shifts like that of aging population, and the ability to accommodate population-specific transportation modes. That fi e-tuning effort mainly aimed to address individual- level attributes of age and gender through the means of a greedy-heuristic data- fitting algorithm that was implemented in the matrix programming language GAUSS. Validation of Srinivasan et al. [9] algorithm yielded satisfactory distributions of household, size, age, gender, and employment status; however, the distributions for all other variables did not match well.

Given the limited number of attributes that can be synthesized per agent, researchers had to further improve the IPF approach to overcome this limitation. Pritchard and Miller [10] introduced a method that implements IPF approach with sparse list-based data structure that allows more attributes per agent. Additionally, they used both the conventional Monte Carlo integerization procedure and the conditional Monte Carlo to synthesize a list of individual agents from fitted tables. Despite their thorough efforts, the study of Pritchard and Miller [10] had only a minor impact on goodness-of-fit, relative to the conventional approach.

Auld and Mohammadian [11] developed a methodology to improve the basic IPF population synthesis routine in a manner that accounts for multiple levels of analysis units—control variables, which was a limitation to the population synthesizers mentioned hereinabove. Their methodology, named multilevel control, allows population characteristics to be replicated for multilevel synthetic population with one level (such as households) serving as the base level of analysis. After a runtime of 16 hours, the proposed method was able to synthesize a 7.9 million agent population for Chicago, IL, with an improved fit of the synthesized individual-level characteristics when compared with synthesis procedures that do not account for individual-level controls. The study concluded that the improved fit comes at no cost to the fit against household-level controls. However, the developed methodology was never experimented as to synthesizing commercial- or business-related agents.

Lee and Fu [12] realized that the IPF-based population synthesis approaches, specifi ally the original synthetic reconstruction method [6] and the complimentary combinatorial optimization method [13], are not generally applicable to all population synthesis scenarios. Based on a comparison by Ryan et al. [14], Lee and Fu [12] concluded that combinatorial optimization method produces more accurate demographic information for populations over a small area and that the population synthesis problem should be evaluated from an optimization point of view. In addition, they explored how the estimation of a multiway demographic table can be formulated and solved as a constrained optimization problem in full consideration of both household- and individual-level attributes. Accordingly, that study tackled the inconsistency problem through an approach that is based on the minimum cross-entropy theory. The validity of that model was confi med through a case study in Singapore, through which results from a 10,641 household study area were superior to conventional IPF approaches. However, Lee and Fu [12] did not provide a full-scale application which constrains the applicability of their model to theoretical applications only.

Zhu and Ferreira [15] were intrigued by the inability of the standard IPF algorithm to fit marginal constraints on multiple agent types simultaneously. Hence, they developed a two-stage population synthesizer that utilized IPF on the fi st stage and then estimated the spatial pattern of household-level attributes through a second stage IPF-based approach. Their two-stage algorithm consisted of four distinctive steps. The fi st step involved developing an estimate joint distribution of household- and individual-level attributes. In the second step, households and individuals were drawn from microdata samples. The third step consisted of a conventional IPF with household type and parcel capacity marginal constraints. The fourth and last step included an estimated marginal distribution of other attributes from the fitted model. To validate their approach, Zhu and Ferreira [15] generated synthetic population for Singapore. Their evaluation approach involved four comparisons, namely, fitting only for households-level constraints, fitting for both household- and individual-level constraints, allocating households to buildings while constraining building capacity, and repeating the previous comparison with income level constrained. Validation results yielded realistic spatial heterogeneity while preserving some of the joint distribution of household and locational characteristics.

Choupani and Mamdoohi [16] addressed the issue of integerization of IPF results in non-integer values instead of integers, for example, fractions of house- hold- or individual-level attributes for zones. In doing so, they proposed a binary linear programming model for tabular rounding in which the integerized table totals and marginals perfectly fit to input data obtained from the Census Bureau. The main advantages of using tabular rounding were that it did not bias joint or marginal distributions of socioeconomic attributes of minority demographic groups and it minimized the distortion to the correlation structure of household- and individual-level non-integer tables. Furthermore, the tabular rounding approach outperformed all other eight rounding approaches. In addition, sensitivity analysis of tabular rounding demonstrated that small and large values are equally signifi ant when it comes to integerization. Their fi dings were confi med by a comprehensive literature review [17] that they performed 1 year later, which concluded that IPF is the most feasible approach for synthesizing populations for agent- and activity- based transportation models, once integer conversion and zero-cell issues were resolved. In addition, they confi med that tabular rounding is the most effici t and feasible solution for the integerization issue.

Most recently, in an effort to further enhance the IPF approach, Otani et al. [18] identifi d an issue that they named the modifiable attribute cell problem (MACP) which arises from combining discrete categories of individual-level attributes or due to the contiguous nature of those attributes. The proposed solution to the MACP issue was identifi d as "the organized cell set" which is the best combina- tion of categories. The procedure to identify the best organized cell set consists of five steps. The fi st step involves aggregation of the elemental cell set to fi d several cases of cell organization that generate large cells. The second step involves constructing base-year data using the conventional IPF approach. The third step focuses on forecasting using microscopic simulation. The fourth step involves identifying the statistically acceptable cell value using a Student's t-test. The fi h and fi al step involves considering the case with minimum number of cells to be the best cell organization. Th s method is computationally complex and cannot be performed using conventional optimization algorithms. Yet, it is the sole identifi- ble solution to the modifiable attribute cell problem.

Iterative proportional updating approach

The iterative proportional updating approach is a heuristic approach that was developed by Ye et al. [19] to address the drawbacks of the IPF approach. Specifi ally, the IPU approach addresses the issue of control for individual-level attributes and joint distributions of personal characteristics. The IPU algorithm matches both household- and individual-level attributes in a computationally effici t manner by iteratively adjusting and reallocating weights among households of a specific type until both household- and individual-level attributes are matched. Another advantage of the IPU approach is its practicality from the implementation and computational points of view. Eq. (2) represents the mathematical optimization problem as addressed by the IPU approach. In addition, the IPU approach has been generally described in 23 computational steps that can be easily coded in most, if not all, programming languages:

$$\text{Minimize} \sum_j \left(\frac{\sum_i d_{i,j} w_i - c_j}{c_j}\right)^2 \text{ or } \sum_j \frac{\left(\sum_i d_{i,j} w_i - c_j\right)^2}{c_j} \text{ or } \sum_j \frac{\left|\sum_i d_{i,j} w_i - c_j\right|}{c_j} \tag{2}$$

Subject to $w_i \geq 0$

where i, denotes a household ($i = 1, 2, \ldots, n$); j, denotes the constraint or population characteristic of interest ($j = 1, 2, \ldots, m$); $d_{i,j}$, represents the frequency of the population characteristic (household/person type j in household i); w_i, is the weight attributed to the ith household; c_j, is the value of the population characteristic j.

Furthermore, Ye et al. [19] proposed an alternative method to address the zero-cell problem that undermined the IPF practicality. Their method is based on borrowing the prior information for the zero cells from PUMS data for the entire region, where zero cells are not likely to exist as long as the control variables of interest and their categories are defi ed appropriately. However, that method has the inherent risk of overrepresenting the demographic group of interest. Despite their attempt to overcome the zero-cell problem, the researchers could not over- come the zero-marginal problem that may result due to nonexistence of a certain attribute in households of a certain geographic area, for example, having no low- income households in a certain census block or tract. Furthermore, a review by Müller and Axhausen [3] pointed to the lack of a theoretical proof of convergence.

Combinatorial optimization approach

The combinatorial optimization approach was materialized by Abraham et al. [20] and is a versatile approach capable of matching targets at multiple agent levels for both household- and individual-level attributes. A combinatorial optimiza- tion approach is generally simpler and more direct than IPF. Mostly, it starts by the creation of a trial population from the disaggregate sample data, and then the overall level of fit is assessed across all marginal targets. Units from the trial population are swapped with units chosen from the disaggregate samples, and when the measure of fit improves, the swap is made. Th s is implemented through a proprietary computer program that fi st identifies a list of units whose aggregate attribute values match a pre-specifi d set of corresponding target values and then iteratively performs one of three operations, name- ly, adding a unit from the sample to the list, subtracting a unit, or swapping a unit between the sample and the previously iden- tifi d list. That process is performed on a zone-by-zone level with equal probability of the three actions (i.e., add, subtract, or swap) being considered. The developed algorithm was applied to

California and Oregon to synthesize populations for their models. The California application served the California Statewide Travel Demand Model including short- and long-distance travel considering personal and commercial vehicles. The Oregon application served the Oregon Statewide Integrated Model, which included employment synthesis for 34 industries. Both model applications resulted in a near-perfect fit for synthesized populations. Generally, the population synthesis procedure using combinatorial optimization has proven to be fast and flex ble with the possibility for application to both households and employment scenarios. However, this algorithm can be further improved by using multicore and parallel computing techniques.

Markov process-based approaches

As demonstrated, hereinabove, IPF, IPU, and combinatorial optimization approaches rely on cloning attributes that were captured in microdata. In addition, they all share key drawbacks including (a) fitting of a contingency table while ignoring other solutions matching the available data; (b) loss of heterogeneity that has been captured in the microdata due to cloning rather than true population synthesis; (c) dependency on the accuracy of captured data to determine the cloning weights which may replicate inherent inaccuracies; and (d) limited scalability, in terms of the number of attributes of synthesized agents. Hence, Markov process-based approaches were developed to overcome such drawbacks and to offer an approach that truly synthesizes populations instead of cloning them.

The earliest notable effort in this direction was pioneered in 2013 by Farooq et al. [21] who developed a Markov chain Monte Carlo (MCMC) simulation-based approach for synthesizing populations. The proposed approach is a computer- based simulation technique that can be used to simulate a dependent sequence of random draws from complicated stochastic models. To synthesize populations that approach uses three sources of data, namely, (a) zoning systems such as census blocks, census tracts, counties, and states; (b) sample of individuals such as the North American PUMS and the European Sample of Anonymized Records (SARs); and (c) cross-classifi ation tables for socioeconomics and demographics like income by age at a certain zoning level. Assuming that in a given spatial region at any point in time there exists a true population, the MCMC simulation-based approach synthesizes that population by drawing the individual attributes from their uniquely joint distribution using the available partial views while ensuring that the empirical distribution in the synthetic population is as close as possible to the unique actual distribution of that population. The proposed approach was applied to the Swiss census data, and results were compared against those developed by a conventional IPF approach. Eq. (3) illustrates the standardized root mean square error (SRMSE)-based goodness-of-fit tests that were performed on each case, and results indicated that MCMC simulation-based synthesis outperformed IPF synthesis while featuring a higher level of heterogeneity:

$$SRMSE = \frac{\left(\sum_{i=1}^{m}\cdots\sum_{j=1}^{n}(R_{i..j} - T_{i..j})^2/N\right)^{1/2}}{\sum_{i=1}^{m}\cdots\sum_{j=1}^{n}T_{i..j}/N} \tag{3}$$

where N, is the total number of agents; $R_{i...j}$, is the number of agents with attribute values $_{i...j}$ in the population synthesized; $T_{i...j}$, is the number of agents with attribute values $_{i...j}$ in the actual population.

Two years later, in 2015, Casati et al. [22] proposed an extension of the MCMC simulation-based approach to simultaneously combine both individual- and house- hold-level attributes in a process that was named hierarchical MCMC. Furthermore, generalized raking was introduced as a technique to fit the simulated synthetic population to actual observed control totals. The hierarchical MCMC is a combination of two methods: (a) an extension of the original MCMC method that allows producing hierarchies of persons grouped into households and (b) a post-processing method to satisfy known control totals on both the individualand household- level. That extension aimed to synthesize populations with a hierarchical structure that is based upon ordering the agents living in the same household according to their household roles. The general formulation of the extension is based upon the defin tion of three groups of agent types (viz., owners, intermediate, and others) running Gibbs sampling on the three groups and merging subpopulations. The proposed approach was applied to the 2008 household interview travel survey of Singapore. The application resulted in realistic synthetic populations, and SRMSE- based test confi med the goodness-of-fit of synthesized populations and their generated hierarchical structures.

Saadi et al. [23] proposed an integrated MCMC approach and profiling-based methods to capture the behavioral complexity and heterogeneity of synthesized agents. Th s approach used two types of datasets, namely, (a) aggregated sociode- mographic and transportation-related variables derived from household travel surveys and (b) individual activity-travel diaries collected from travel diary surveys. The integrated approach consists of six steps that run on those two data types. The fi st step involves performing a MCMC simulation on the sociodemographic dataset. The second step concerns synthesizing population by a Gibbs sampling procedure. The third step selects sociodemographics to compare behaviors in the activity-travel patterns. The fourth step uses results from the previous two steps to cluster synthesized populations according to sociodemographics and related activity sequences. The fi h step utilizes multiple sequence alignments to estimate hidden Markov model (HMM). The fi al step characterizes clusters including mixed socioeconomic effects. The integrated approach was applied to the 2010 Belgian household daily travel survey. Results indicated that the integrated approach effectively captured the behavioral heterogeneity of travelers. In addition, comparisons against IPF and IPU approaches demonstrated that the proposed integrated approach is adequately adapted to meeting the demand for large-scale microsimulation scenarios of urban transportation systems.

Realizing the advantages of Markov process-based approaches, Saadi et al. [24] developed an extended HMM-based approach which promised better alternatives than the existing ones. More specifi ally, the proposed HMM-based approach promised great flex bility and effici cy in terms of data preparation and model training while being able to reproduce the structural configur tion of a given population from an unlimited number of micro samples and a marginal distribution. The HMM-based approach considers population synthesis as a variant of the standard decoding problem, at which the state sequences are supposed to be unknown. Accordingly, the maximum likelihood estimators related to the transition states were determined through the Viterbi algorithm. An important advantage of the HMM-based approach is its ability to handle both continuous and discrete variables, which addresses the inherent issue of loss of information due to aggregation of continuous variables like age. Also, the proposed HMM-based approach satisfies the need to discretize continuous variables to meet the fundamental limitation of Markov process to discrete states. The statistical and machine Learning Toolbox of MATLAB was used to generate sequences from an estimated HMM that were applied to the 2013 Belgian National household travel survey. Th ee simulations were run to illustrate the HMM-based approach. The fi st simulation tested the combined effects of scalability and dimensionality. The second simulation compared the HMM-based approach against IPF, and the third demonstrated the advantage of the HMM-based approach over IPF using various samples. Simulation results indicated that the proposed HMM-based approach provided accurate results due to its ability to reproduce the marginal distributions and their corresponding multivariate joint distributions with an acceptable error. Furthermore, the HMM-based approach outperformed IPF for small sample sizes while using smaller amount of input data than IPF. In addition, simulation results demonstrated that the HMM-based approach can integrate information provided by several data sources to allow good estimates of synthesized population.

Fitness-based synthesis approach

To address the inability of the IPF approach to deal with multilevel controls, Ma and Srinivasan [25] developed the fitness-based synthesis approach that directly generates a list of households to match several multilevel controls without the need for determining a joint multiway distribution. The FBS approach generally involves selecting a set of households from the seed data, like PUMS, such that tract-level controls are satisfi d. The FBS approach starts with an initial set of households that can either be a null set or a random sample from the seed data. Then, the population of each census tract is synthesized in an iterative fashion, with one household being either added or removed from the current list in each iteration. Count tables, defi ed in terms of control attributes, are used to track the number of households of each type that have already been included. The FBS approach implements an adding or removing procedure, while swapping is not considered. The main criteria in the FBS approach is the reduced sum of squared error for addition F^{in}_{I} and corresponding error for removal F^{in}_{II} as illustrated by Eqs. (4) and (5):

$$F_I^{in} = \sum_{j=1}^{J}\sum_{k=1}^{K_j}\left[\left(R_{jk}^{n-1}\right)^2 - \left(R_{jk}^{n-1} - HT_{jk}^{i}\right)^2\right] \tag{4}$$

$$F_{II}^{in} = \sum_{j=1}^{J}\sum_{k=1}^{K_j}\left[\left(R_{jk}^{n-1}\right)^2 - \left(R_{jk}^{n-1} + HT_{jk}^{i}\right)^2\right] \tag{5}$$

Subject to $F_I^{in} + F_{II}^{in} = -2\sum_{j=1}^{J}\sum_{k=1}^{k_j}\left(HT_{jk}^{i}\right)^2$

where $R^{n}{}_{JK}^{-1} = T_{jk} - CT^{n}{}_{JK}^{-1}$; j, is an index representing the control (and the corresponding count) tables; J, is the total number of control (or count) tables; jk, is an index representing the different cells in a table; Tjk, represents the value of cell k in control table j; $CT^{n}{}_{jk}^{-1}$, represents the value of cell k in count table j after itera tion $n-1$; $R^{n}{}_{jk}^{-1}$, is the number of households/persons required to satisfy the target for cell k in control table j after iteration $n-1$; $HT^{i}{}_{jk}$, is the contribution of the ith household in the seed data to the kth cell in control table j.

Th ee applications of the FBS approach were performed to demonstrate the feasibility of incorporating many controls at multiple levels in the synthesis and increased accuracy of synthesized population. The three applications were performed using the 2000 Census data for 12 census tracts in Florida. The fi st application involved population synthesis using the IPF approach with only house- hold-level controls. The second application involved population synthesis using the proposed FBS approach with few household- and individual-level controls. The third application also involved population synthesis using the FBS approach but with signifi antly larger number of controls. Validation for the three applications was performed by comparing the mean absolute error against 22 artific al census tracts that were created by randomly selecting subsets of households from the 2000 PUMS. Validation results demonstrated that FBS outperformed IPF and demonstrated effici cy and scalability. In addition, FBS did not require many iterations as it required only one to three times the number of households to be synthesized. In addition, the proposed FBS approach addresses the notorious IPF issues of zero-cell problems, computational resources (memory), and non-integers cell value in the joint-distribution tables.

Hafezi and Habib [26] refi ed the FBS approach, and the refi ed FBS population synthesizer was examined by three models. The fi st model used household- level control tables. The second model used individual- and household-level control tables, and the third model used weighting individual-and household-level control tables. The models were applied to the province of Nova Scotia in Atlantic, Canada, using the 2006 Canadian

Census and Public Use Microdata File (PUMF). The refi ed approach was implemented using the sparse matrix technique package in MATLAB that is based on high-level matrix programming for numerical computation. The three models were validated by error percentages and goodness- of-fit evaluation. Validation results indicated that the refi ed FBS approach can effici tly obtain a satisfactory result using both individualand household-level control tables. However, higher homogeneity was achieved within the third model.

Emerging approaches

Other emerging approaches have been developed in an attempt to replace the IPF approach or to overcome one or more of its drawbacks. Emerging approaches include Bayesian network, annealing algorithm, linear programming, heuristicbased, copula-based, and entropy maximization approaches. The following paragraphs introduce each of the emerging approaches.

The Bayesian network approach was developed by Sun and Erath [27] in 2015. The proposed Bayesian network approach is a probabilistic population synthesizer that is intended as an alternative to approximate the inherent joint distribution in a more effici t manner. Using a graphical model, the proposed Bayesian network approach encodes probabilistic relationships, like causality or dependence, among a set of variables. The advantages of Bayesian network models lie in their ability to learn the structure of population systems, particularly when the number of attributes of interest is large using limited amounts of microdata. The Bayesian network approach was founded on the inference of the joint distribution—that is, perceiving the population synthesis problem as an inference of a multivariate probability distribution of demographic and socioeconomic household- and individual-level attributes. Like the Markov process-based approaches, the Bayesian network approach does not require marginals as input. In addition, it does not require any conditionals since structure learning and parameter estimation are inherently integrated in the learning model. The performance of the proposed Bayesian network approach was demonstrated through an application to the 2010 household interview travel survey of Singapore. The Bayesian network approach demonstrated good performance as illustrated by low SRMSE values. It also demonstrated good heterogeneity in synthetic population when the size of PUMS is less than 70% of the full population.

The simulated annealing (SA) algorithm was developed by Kim and Lee [28] to synthesize populations for activity-based models. The proposed SA algorithm is built upon the concepts of thermodynamics and metallurgy and was fi st introduced as a generic heuristic method for discrete optimization. The Metropolis-Hastings Algorithm was employed to solve the inherent problems of hill climbing and cooling schedule when applying SA to population synthesis. The proposed algorithm consists of seven steps. The fi st step concerns setting the maximum number of iterations. The second step sets up the total amount of columns and rows in the population and enters observed values of sample distribution. The third step sets up the before-distribution, which is composed by random numbers, while satisfying the total amount of restrictive conditions. The fourth step sets up the after-distribution, which is also composed by random numbers that satisfy total amount restrictive conditions. The fi h step involves calculation of absolute error on the before-/ after-distributions as well as observed data. The sixth step involves calculation of selection probability. The seventh and fi al step iterates steps 4 through 6 and ends the calculations when the absolute error (calculated in the fi h step) has the smallest value or satisfies ending conditions. The SA algorithm was implemented using the household travel diary survey from the Korean National Statistics Offi . Results from the implementation indicated the need for further verifi ation of the accuracy of this algorithm.

The linear programming (LP) approach was developed by Vovsha et al. [29] to synthesize populations as part of an activity-based model developed for the Maricopa Association of Governments. The LP approach is an analytical method that balances a list or sample of household weights to meet the controls imposed at some spatial level, typically, for each traffi analysis zone (TAZ). Features of the LP approach include (a) the general formulation of convergence of the balancing procedure with imperfect controls, (b) optimized discretization of weights while preserving the best possible match to the controls, and (c) ability to set controls at multiple spatial levels. In addition, the proposed LP approach featured an innovative discretizing method applied for the household weights and integrated with the balancing procedure. While validation of the proposed LP approach is questionable, it still demonstrates reasonable accommodation to various fi e-resolution spatial levels that are much needed by newer-generation activity- and agent-based models.

The heuristic-based approach was developed by Zhuge et al. [30] to address two IPF limitations that received less attention from earlier studies. The fi st limitation stems from the existence of various solutions for one target marginal distribution. The second limitation stems from the optimization nature of population synthesis with the objective function being minimizing the mean absolute percentage error (MAPE) of control variables. The proposed heuristic-based approach consists of 11 steps arranged in three parts. The fi st part, including steps 1 and 2, is used to generate the initial household weights. The second part, including steps 3 through 11, adjusts the household weights until a stop criterion is met. The third part, including steps 10 and 11, calculates the adjustment steps and adjustment range, which are two fundamental parameters of the approach. The 2007 household travel survey data from Baoding, China, were used as a case study. Results indicated that heuristic-based approach cannot perform as well as IPF-based on comparing MAPE values for both approaches.

Most recently, the copula-based approach was proposed by Kao et al. [31] to address previously identifi d limitations of IPF approach. Copulas are joint probability distributions with uniform marginal, which are a relatively new statistical tool. Hence, the copula-based approach was designed to preserve marginal distributions and dependence structure between variables. The proposed method was tested for the state of Iowa, and the results were compared with the IPF approach using mean, median, and correlation matrices. The synthesized households resulted in the same local statistics at each block group, but having similar intervariable correlations as described in the PUMS suggests the applicability of the copula-based approach.

Another recent effort to develop an alternative to IPF approaches resulted in the development of entropy maximization-based population synthesizer by Paul et al. [32] which handles multiple geographies and avoids algorithmic errors. The entropy maximization approach was developed as part of the Oregon Department of Transportation (ODOT) effort to utilize an open-source population synthesis platform. The approach consists mainly of two algorithms. The fi st algorithm, namely, list balancing, fi ds weights that match the given marginal control distributions. The second algorithm, namely, integerizing, implements a LP-based procedure to covert fractional weights to integers. The proposed entropy maximization-based approach was implemented in Python and made heavy use of the Pandas and NumPy libraries, which allow for vectorization of operations to reduce overall runtime. Validation results against those of IPF approach were promising and demonstrated reasonable match to controls.

Conclusion

Th s study presented a critical, comprehensive literature review of population synthesizers starting from the early efforts through the most recent approaches. The review and synthesis indicated that, despite its identifi d limitations and draw-backs, IPF approach is the most feasible and widely used population synthesizer. All other studies and efforts used it as a reference for comparison and produced similar or slightly improved results. Evidently, IPF has its drawbacks and limitations. Yet reviewed literature indicates that there is no single approach that can result in an effici t and accurate population synthesizer. However, an integration of robust methods appears as the most promising approach, like the effort of Fournier et al. [33] where the limitations of IPF are resolved by combining five methods into an integral framework for population synthesis. **Table 1**, in the *Supplemental Information* section, summarizes the advantages and disadvantages of the presented approaches.

Almost three-decade old, yet the IPF approach is still being used in state-of-the- art simulation platforms like MATSim. Given that IPF is the most studied approach and the fact that none of the alternatives provided an out-of-the-box solution, IPF is preferred approach by modelers and practitioners. Th s conclusion is confi med by the fi dings of Saadi et al. [34], who investigated the influence of scalability on the accuracy of different population synthesizers using both fitting- and generation- based approaches. Their results revealed that simulation-based approaches are more stable than IPF approaches when the number of attributes increases; however, IPF approaches are less sensitive to changes in sample size.

Approach	Advantage(s)	Disadvantage(s)
Iterative proportional fi ting (IPF)	Synthesized estimates maintain the same odds ratios as those in the sample table Most studied and improved approach with more than 20 years of continuous refinements Widely available with ready-to-use implementations in several computer programming languages	Does not provide an answer to the zero-cell problem in the public use microdata sample (PUMS) Unable to control for statistical distributions of both household- and individual-level attributes Limited number of attributes that can be synthesized per agent
Iterative proportional updating (IPU)	Addresses the issue of control for individual-level attributes and joint distributions of personal characteristics Computationally effici t Described in 23 computational steps that can be easily coded in most programming languages	Cannot overcome the zero-marginal problem that may result due to nonexistence of a certain attribute in the households of a certain geographic area
Combinatorial optimization	Generally simpler and more direct than IPF Fast and flex ble with the possibility for application to both households and employment scenarios	Implementation is limited to a proprietary computer program Resource-demanding and needs multicore, parallel computers
Markovprocess-based approach	Truly synthesizes populations instead of cloning them Meets the demand for large-scale microsimulation scenarios Can handle both continuous and discrete variables	Requires extensive knowledge of computer programming Difficult to trace errors Refinement for specific scenarios or locations requires substantial redevelopment of the computer algorithm
Fitness-based synthesis (FBS)	No need for determining a joint multiway distribution Addresses the notorious IPF issues of zero-cell problems	Requires extensive knowledge of the sparse matrix technique package in MATLAB that is based on high-level matrix programming for numerical computation
Emerging	Scalable and adaptive	Requires advanced expertise in Python

approaches	Addresses all disadvantages of IPF approach	and makes heavy use of the Pandas and NumPylibraries Limited successful applications compared to IPF

Table 1.

Key advantages and disadvantages of population synthesis approaches.

Overall, this study provides a critical review and comprehensive synthesis of population synthesis approaches that can serve as a valuable reference to future efforts focusing on population synthesis for activity- and agent-based transportation models.

Acknowledgements

Th s study was performed in support of the project "Technology Influence on Travel Demand and Behaviors" that was sponsored by the US Department of Transportation Offic of the Assistant Secretary for Research and Technology (OST-R) through the Southeastern Transportation Research, Innovation, Development, and Education (STRIDE) Center under Contract No. 69A3551747104 with matching funds from the Alabama Department of Transportation (ALDOT).

Author details

Ossama E. Ramadan and Virginia P. Sisiopiku*
University of Alabama at Birmingham, Birmingham, AL, USA

*Address all correspondence to: vsisiopi@uab.edu

References

[1] Axhausen KW, Gärling T. Activity- based approaches to travel analysis: Conceptual frameworks, models, and research problems. Transport Reviews. 1992;**12**(4):323-341

[2] Bowman JL, Rousseau G, editors. Validation of Atlanta, Georgia, regional commission population synthesizer. In: Innovations in Travel Demand Modeling Conference; 2006; Austin, TX. Washington, DC: Transportation Research Board; 2008

[3] Müller K, Axhausen KW, editors. Population synthesis for microsimulation: State of the art. In: Transportation Research Board 90th Annual Meeting. Washington, DC: Transportation Research Board; 2011

[4] Deming WE, Stephan FF. On a least squares adjustment of a sampled frequency table when the expected marginal totals are known. The Annals of Mathematical Statistics. 1940;**11**(4):427-444

[5] Papacostas CS, Prevedouros PD. Transportation Engineering and Planning. 3rd ed. Englewood Cliffs, NJ: Prentice Hall; 2001

[6] Beckman RJ, Baggerly KA, McKay MD. Creating synthetic baseline populations. Transportation Research Part A: Policy and Practice. 1996;**30**(6):415-429

[7] Arentze T, Timmermans H, Hofman F. Creating synthetic household populations: Problems and approach. Transportation Research Record. 2014;**2007**:85-91

[8] Guo JY, Bhat CR. Population synthesis for microsimulating travel behavior. Transportation Research Record. 2014;**2007**:92-101

[9] Srinivasan S, Ma L, Yathindra K. Procedure for Forecasting Household Characteristics for Input to Travel- Demand Models. Report No. TRC- FDOT-64011-2008. Tallahassee, FL: Florida Department of Transportation; 2008

[10] Pritchard DR, Miller EJ, editors. Advances in agent population synthesis and application in an integrated land use and transportation model. In: Transportation Research Board 88th Annual Meeting. Washington, DC: Transportation Research Board; 2009

[11] Auld J, Mohammadian A. Effici t methodology for generating synthetic populations with multiple control levels. Transportation Research Record. 2010;**2175**:138-147

[12] Lee D-H, Fu Y. Cross-entropy optimization model for population synthesis in activity-based microsimulation models. Transportation Research Record. 2011;**2255**:20-27

[13] Williamson P, Birkin M, Rees PH. The estimation of population microdata by using data from small area statistics and samples of anonymised records. Environment and Planning A: Economy and Space. 1998;**30**(5):785-816

[14] Ryan J, Maoh H, Kanaroglou P. Population synthesis: Comparing the major techniques using a small, complete population of fi ms. Geographical Analysis. 2009;**41**(2):181-203

[15] Zhu Y, Ferreira J. Synthetic population generation at disaggregated spatial scales for land use and transportation microsimulation. Transportation Research Record. 2014;**2429**:168-177

[16] Choupani A-A, Mamdoohi AR. Population synthesis in activity- based models. Transportation Research Record. 2015;**2493**:1-10

[17] Choupani A-A, Mamdoohi AR. Population synthesis using iterative proportional fitting (IPF): A review and future research. Transportation Research Procedia. 2016;**17**:223-233

[18] Otani N, Sugiki N, Vichiensan V, Miyamoto K. Modifiable attribute cell problem and solution method for population synthesis in land use microsimulation. Transportation Research Record. 2018;**2302**:157-163

[19] Ye X, Konduri KC, Pendyala RM, Sana B, Waddell P, editors. Methodology to match distributions of both household and person attributes in generation of synthetic populations. In: Transportation Research Board 88th Annual Meeting. Washington, DC: Transportation Research Board; 2009

[20] Abraham JE, Stefan KJ, Hunt JD, editors. Population synthesis using combinatorial optimization at multiple levels. In: Transportation Research Board 91st Annual Meeting. Washington, DC: Transportation Research Board; 2012

[21] Farooq B, Bierlaire M, Hurtubia R, Flötteröd G. Simulation based population synthesis. Transportation Research Part B: Methodological. 2013;**58**:243-263

[22] Casati D, Müller K, Fourie PJ, Erath A, Axhausen KW. Synthetic population generation by combining a hierarchical, simulation-based approach with reweighting by generalized raking. Transportation Research Record. 2015;**2493**:107-116

[23] Saadi I, Mustafa A, Teller J, Cools M. Forecasting travel behavior using Markov chains-based approaches. Transportation Research Part C: Emerging Technologies. 2016;**69**:402-417

[24] Saadi I, Mustafa A, Teller J, Farooq B, Cools M. Hidden Markov model-based population synthesis. Transportation Research Part B:

Methodological. 2016;**90**:1-21

[25] Ma L, Srinivasan S. Synthetic population generation with multilevel controls: A fitness-based synthesis approach and validations. Computer- Aided Civil and Infrastructure Engineering. 2015;**30**(2):135-150

[26] Hafezi MH, Habib MA, editors. Synthesizing population for agent- based microsimulation modeling in Atlantic Canada. In: Transportation Research Board 94th Annual Meeting. Washington, DC: Transportation Research Board; 2015

[27] Sun L, Erath A. A Bayesian network approach for population synthesis. Transportation Research Part C: Emerging Technologies. 2015;**61**:49-62

[28] Kim J, Lee S. A simulated annealing algorithm for the creation of synthetic population in activity-based travel demand model. KSCE Journal of Civil Engineering. 2015;**20**(6):2513-2523

[29] Vovsha P, Hicks JE, Paul BM, Livshits V, Maneva P, Jeon K, editors. New features of population synthesis. In: Transportation Research Board 94th Annual Meeting. Washington, DC; 2015

[30] Zhuge C, Li X, Ku C-A, Gao J, Zhang H. A heuristic-based population synthesis method for micro-simulation in transportation. KSCE Journal of Civil Engineering. 2017;**21**(6):2373-2383

[31] Kao S-C, Kim HK, Liu C, Cui X, Bhaduri BL. Dependence-preserving approach to synthesizing household characteristics. Transportation Research Record. 2018;**2302**:192-200

[32] Paul BM, Doyle J, Stabler B, Freedman J, Bettinardi A, editors. Multi-level population synthesis using entropy maximization-based simultaneous list balancing. In: Transportation Research Board 97th Annual Meeting. Washington, DC: Transportation Research Board; 2018

[33] Fournier N, Christofa E, Akkinepally AP, Azevedo CL, editors. An integration of population synthesis methods for agent-based microsimulation. In: Transportation Research Board 97th Annual Meeting. Washington, DC; 2018

[34] Saadi I, Eftekhar H, Teller J, Cools M, editors. Investigating the scalability in population synthesis: A comparative approach. In: Transportation Research Board 96th Annual Meeting. Washington, DC: Transportation Research Board; 2017

6

A Multitiered Holistic Approach to Traffic Safety: Educating Children, Novice Teen Drivers and Parents, and Crash Investigators to Reduce Roadway Crashes - An Eight-Year Introspective Project

Claudia M. Knezek, Susan Polirstok, Roxie James, Anthony Pittman and Gary Poedubicky

Abstract

The Strategic Prevention Framework (SPF), developed by the Substance Abuse and Mental Health Services Administration (SAMHSA), was adopted as an effective injury reduction model for reducing driver injury crashes on community roadways. Kean University and the New Jersey Division of Highway Traffi Safety (NJDHTS) adopted crash prevention strategies involving education and enforcement outreach. First, an effective K-12 traffi safety program was established for supporting driver education training and then crash investigation training, and a statewide traffi safety specialist certifi ation was promoted within the law enforcement community. Th s successful outreach initiative also involves community representation, including law enforcement personnel and parents of novice drivers. Best practices have been established in New Jersey by four traffi safety specialist (TSS)-Level 2 leaders, with over 100 more TSS offic s waiting to qualify for this second tier. Future plans involve an outreach program for offic s to develop traffi safety programs in their communities while qualifying for the TSS-Level 2 designation.

Keywords: traffic afety, novice driver, crash investigation

Introduction

The Haddon matrix is a public safety model that was developed in 1980 to further standardize safety analysis. The matrix is a two-dimensional model that adapts principles of public health to crash injuries and fatalities. The fi st domain of the model involves rows that are divided into pre-crash, crash, and post-crash phases. The second domain is known as influencing factors of injury: human, vehicle/equipment, physical environment, and socioeconomic. The Matrix has been successfully adopted to evaluate crash sites and/or related conditions. Results often provide information on safety issues and potential solutions. Th s planning tool is effective for establishing countermeasures based on crash-related data collection and collaborations with involved agencies. The countermeasures offer a variety of solutions addressed through education, enforcement, engineering, and emergency response solutions (the 4 Es of Safety) [1].

In 2005, John Hopkins School of Public Health had successfully adopted the Haddon matrix in preparation for a public health threat (e.g., SARS outbreak and dirty bomb response). Influencing factors were effectively adapted to facilitate the conditions surrounding emergency readiness [1]. Specifi ally, the "human" category was renamed "host," and "vehicle/equipment" category had been changed to "agent/vehicle." Otherwise both models were comparable in analysis. Th s phase factor approach offered a multiple factor concept for delivering public health interventions with strategies to prevent, respond to, or mitigate injuries. Pre-event activities include risk assessment, communications, and prevention efforts, while the event phase deals directly with the crisis. Pre-event activities include risk assessment, communication, community-based medical interventions, counseling, and quarantine measures. Post-event activities involve disaster mitigation, longer-term treatment, risk communication, and recovery efforts.

While the Haddon matrix is an effective tool for addressing pending public health incidents, pre-event strategies are focused on the crisis and not fully involved with prevention. Fortunately, the nationally based Substance Abuse and Mental Health Services Administration (SAMHSA) has successfully adapted the Haddon matrix to include five steps that form a Strategic Prevention Framework (SPF) [2]. Th s new framework offers individual- as well as community-based collaboration for the prevention of crash injuries. Therefore, Haddon environmental strategies of enforcement and education were broadened to include communities in supporting the reduction of roadway crashes. While traditional education remains effective for addressing individual learning needs, communications, public education, social marketing, media advocacy, and media literacy have been used to support crash reductions (e.g., no texting and driving campaigns). After policies and laws are offi ally enacted, supportive enforcement strategies currently involve surveillance, penalties, and consequences for unsafe actions as well as participation in community policing and incentive programs for improving public safety. Lastly, collaboration partnerships support further reduction of motor vehicle injuries and fatalities.

There are two guiding principles of the SPF Framework: understanding the community needs and sustainability in achieving and maintaining long-term results. Once safety issues have been identifi d through analysis of crash data, needed resources are identifi d that include manpower, equipment, and training; a formal plan is then established, as part of the SPF model. The plan is implemented to address population-level challenges. Finally, evaluation involves the ongoing facilitation of prevention efforts with reliance upon a community-based team approach.

In 2010, Kean University and the New Jersey Division of Highway Traffi Safety (NJDHTS) identifi d crash prevention strategies for education and enforcement. First, an effective K-12 traffi safety program was established for supporting driver education training and then crash investigation training, and a statewide traffi safety specialist certifi ation was promoted within the law enforcement community. During the past 8 years, the traffi safety community has worked together with Kean University to support law enforcement and public education-based programs, participate in applied research, develop targeted initiatives, and market K-12 traffi safety educational curricula on behalf of the division. Th s successful outreach initiative also involves community representation, including parents of novice drivers. The following section describes the Strategic Planning Framework as well as the corresponding education and enforcement components.

Traffic safety injury prevention strategic planning framework (SPF) model

Description

In order to support this initiative, two independent advisory committees were formed with some overlap in representation. The ongoing Education Advisory Committee includes membership from the teacher's professional association, known as the New Jersey Association of Health Physical Education Recreation and Dance (NJAHPERD), Driver Education Association representatives, and New Jersey State Department representatives from the Division of Highway Traffi Safety, Motor Vehicle Commission (MVC), and Department of Education (DOE). The Enforcement Advisory Committee collaboration differs with equal levels of state, county, regional, and local representation. Also, the New Jersey State Police are represented, along with leadership from the Police Traffi Offic s' Association of New Jersey (PTOANJ), the New Jersey Division of Highway Traffic fety, and participating universities.

As expected, both committees serve two distinct roles. The Education Advisory Committee is responsible for conducting K-12 traffi safety education including driver education, while the Enforcement Advisory Committee oversees crash investigation training and local traffi safety programs. Fortunately, the traffi safety specialist certifi ation program addresses the ongoing need to educate offic s on new technologies and promote community-based safety projects for college-level credits. **Figure 1** identifies two programs that each advisory committee administers through the injury prevention model. Further information will follow in the next section.

Background

Advisory members are expected to participate in a five-step process to ensure ongoing oversight of the education and enforcement components. Each committee is responsible for (1) assessing the level of crash-related needs and priorities within the state. Then a program component is (2) built to address the targeted need. (3) Strategic plans are developed that involve evidence-based strategies, while (4)

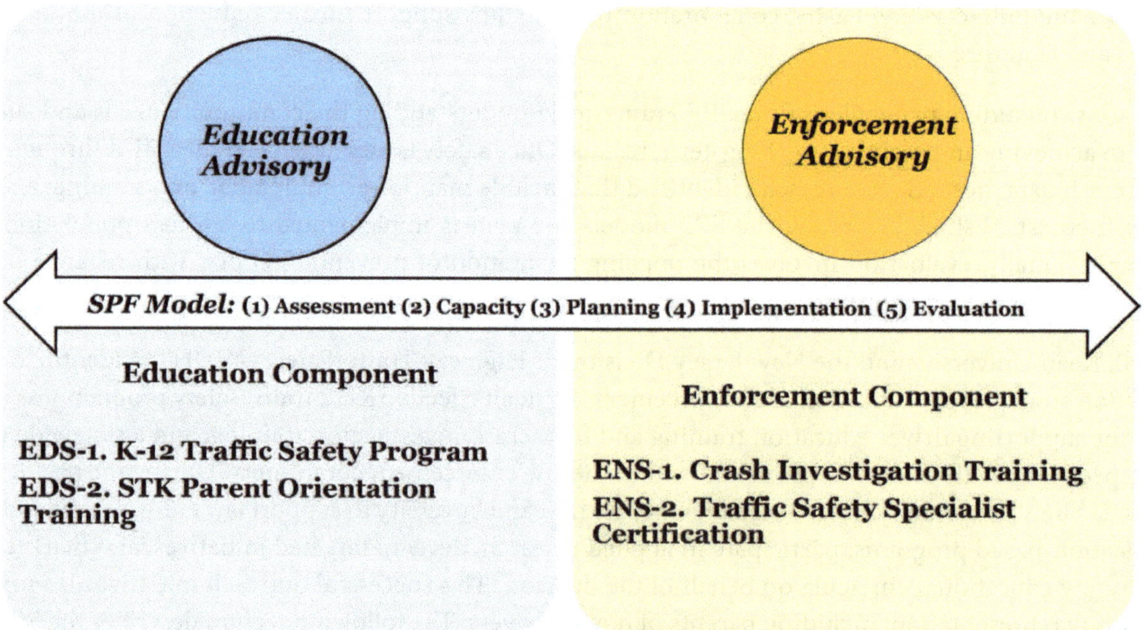

Figure 1.
Traffic safety injury prevention strategic planning framework (SPF).

evidence-based prevention programs are implemented in specifi d areas. Finally, (5) programs are evaluated and monitored for ongoing improvement and success.

The SPF procedure resulted in two very important outcomes for the advisory committees to support. The

Education Committee has targeted their prevention efforts on the development of a K-12 traffi safety component for educating children (e.g., future drivers) on the importance of following traffi safety procedures. Therefore, rules of the road, seat belt usage, and personal safety are practiced before children mature and become novice drivers, when enrolling in driver education course. Since this is a community-based model, students are involved in activities that promote traffi safety in the community. Therefore, parents and children are engaged in establishing a critical "safety" relationship, long before the child becomes a novice driver.

Since the Enforcement Advisory Committee relies on offic s who are dedicated public servants, prevention is focused on implementing community-based safety initiatives to address local needs. Th s approach begins with the participation in extensive crash investigation training, thus enabling offic s to identify potential safety risks that may be overlooked by others in the community. Specifi ally, the traffi safety specialist certifi ation includes office recognition as well as the opportunity to participate in projects that involve crash analysis and adoption of crash reduction countermeasures.

Education component (EDS)

A linear approach to traffi safety has been developed for engaging students in the traffi safety community as a passenger, pedestrian, or bicyclist. Elementary students are trained by health and physical education teachers to adopt safety practices and injury prevention behaviors. Middle school students are then instructed on the assessment and remediation of unsafe traffi conditions, while high school students focus on the role of becoming a driver and safe passenger.

K-12 traffic safety program

Implementation of the K-12 roadway safety curriculum has been in targeted areas of New Jersey, instead of on a statewide basis. The K-12 traffi safety programs include several proven resources which were selected based on their effectiveness (**Table 1**) [9]. A total of 76 lesson plans were developed and selected for bicycle [11], pedestrian [11], traffi safety [10], and supplemental units [8]. A series of developmental skills have been incorporated into the program, beginning with awareness of traffi safety behaviors and fin shing with students serving as peer advocates. Learning goals involve identifi ation of prevention processes, development of strategies for reducing unsafe safety conditions, and assessment of the traffi systems to identify unsafe conditions encountered by at-risk populations. High school students are also required to analyze crash data to further understand and prevent traffi crashes, develop rationales for peer compliance of traffi safety laws, and lead safety advocacy campaigns [12].

The Grade 9–12 section is dominated by driver education training founded upon the MVC Driver Education Manual. As a result, the traffi safety community encountered the need to further promote advocacy and oversight of driver education. When the New Jersey Driver Education Committee (NJ DEC) Charter Plan was adopted in April 2017, this committee began to work on conducting an assessment of classroom instruction and determined that requiring a standardized curriculum would be ineffective, since most current programs are (80%) in compliance with national

Grade level	Cumulative progress indicator (CPI)
Prekindergarten	1. Use safe practices indoors and outdoors (e.g., wear bike helmets, cars are used, and seat belts are worn) 2. Develop an awareness of warning symbols and their meanings (e.g., red light, stop signs, etc.)

Grade	Objectives
Kindergarten-Grade 2	1. Identify ways to prevent injury at home, in school, and in the community (e.g., fi e safety, poison safety, and accident prevention) 2. Identify procedures associated with pedestrian, bicycle, and traffic afety
Grades 3–4	1. Determine the characteristics of safe and unsafe situations, and develop strategies to reduce the risk of injuries at home, in school, and in the community 2. Examine the impact of unsafe behaviors when traveling in vehicles, as pedestrians and when using other modes of transportation
Grades 5–6	1. Summarize the common causes of intentional and unintentional injuries in adolescents and related prevention strategies 2. Summarize the components of the traffic afety system, and explain how people contribute to make the system effective
Grades 7–8	1. Assess the degree of risk in a variety of situations, and identify strategies to reduce intentional and unintentional injuries 2. Analyze the cause of non-compliance with the traffic afety system and their consequences
Grades 9–12	1. Determine the causes and outcomes of intentional and unintentional injuries in adolescents and young adults, and propose prevention strategies 2 Analyze the relationship between alcohol and drug use and the incidence of motor vehicle crashes 3 Develop a rationale to persuade peers to comply with traffi safety laws and avoid detractors 4 Summarize New Jersey motor vehicle laws and regulations, and determine their impact on the incidence of crashes and injuries 5 Plan and implement an advocacy strategy to stimulate action on a state, national, or global health issue, including but not limited to organ/tissue donation

Table 1.
Traffic safety program.

guidelines. Furthermore, the New Jersey model has been ahead of the country with involving parents in novice driver education and will be discussed the next section.

Parent-teen orientation training

A national review of Parent/Teen Graduated Driver Licensing (GDL) programs was conducted in 2011 and yielded limited information on behavioral approaches for parents to teach their teenagers to drive. Fortunately, the Children's Hospital of Philadelphia (CHOP) [3] completed a national study that identifi d teenagers' perspectives on the role of their parents in the GDL process. Under direction of the division, Kean University staff ad developed the Parent/Teen Orientation

Workshop that empowers parents to support their teens in the development of safe driving skills for life. Learning outcomes were incorporated into this 1–1/2-hour orientation program and continue to be measured in pre–/post-study.

The Parent/Teen Orientation Program has been organized into three modules: introduction, practice driving, and enforcement of the graduated driver's license. Th s interactive format requires facilitators to briefly review slides for impact, address talking points on the topic, and engage audiences of parents and teens in discussions. Also, several scenarios are presented as learning experiences (**Table 2**).

Module	Content
Introduction	National/teen driver statistics (i.e. no. 1 killer of teens claiming nearly 6000 lives each year, CHOP) Background on how GDL has worked to improve teen driver safety and save young lives. Parental roles and their impact on teen driver safety
Practice driving	National teen driving statistics (NHTSA) and changes in behavior that can decrease crash risks with experience Strongly encourage parents to expose their young driver to various road and weather conditions noting that it is safer and smarter to expose them to these conditions while you are in the vehicle Discussion of a typical families schedule taking into account the various types of families (one parent, etc.) showing how an hour a week can be found to practice Alternative approaches for teen driving coaches (e.g., family or friends with safe driving habits)
Enforcement of GDL	Presentation of CHOP data showing the safety experience of teens with authoritative parents Teens who describe authoritative parents, as compared to uninvolved parents are: • Half as likely to crash • 71% less likely to drive while intoxicated • 30% less likely to use a cell phone while driving Emphasize that teens need clear instruction and parents act as role models, since they have been teaching their teen to drive by the example since childhood *Practice scenarios:* Exchange between teen and parent when teen's attempt to justify or plead the case for breaking a restriction (curfew-going to a late movie, passenger-carpooling for school activity/sports use of cell phone—they receive a call from parent) Demonstration of the distraction that cell phones create for drivers and the brains inability to focus on both driving and talking on the cell phone at the same time

Table 2.
Parent-teen orientation program.

Enforcement component (ENS)

The crash investigation component establishes baseline knowledge for the enforcement initiatives. Basic investigation involves the development of investigation skills to prevent crash risks and optimize traffic flow when a crash does occur. Recognition and preservation of evidence is accomplished by utilizing photographing techniques, sketching, and measuring vehicle damage. Speed calculations are performed, in order to determine time-distance factors of a crash. Advanced investigation is an analytic approach to measurement, photography, and sequence analysis. Work and energy formulas are used to determine speed, while vectors have been adopted as effective tools for presenting crash analysis findings. This background offers tools for investigators to identify potential crash sites and develop solutions to address potential crash hazards. Further information will be presented that identifies crash investigation training content, the traffic safety specialist requirements, and resultant safety benefits for the community.

Crash investigation program

During the past 20 years, New Jersey Division of Highway Traffi Safety (NJDHTS) cosponsors crash investigation training for municipal, county, and state police personnel. In 2010, Kean University was contracted to work with the New Jersey crash instructors and develop a crash investigation training program. After conducting a national review on crash investigation and locating best fi ld practices, courses were created through the Enforcement Advisory Committee. The curriculum has been based on state needs and national practices. Th s hybrid course design of a traditional and interactive delivery system adequately accommodates the many challenges faced by law enforcement offic s working on the most congested roadways in the nation [4]. A general format of instructional goals and objectives, assessment strategies, content, worksheets, and instructional evaluations has also been employed for the basic and advanced crash investigation courses. These tools have enabled the program to be properly vetted and measured for effectiveness of learning strategies, instructor delivery of content, and ongoing analysis of data collection techniques.

Basic crash investigation

Basic crash investigation is designed to offer police offic s an awareness of skills required to conduct traffi crash investigations. Students are provided with a working knowledge of evidence and information needed upon arrival at the scene. The course addresses how to safely obtain the needed information without further endangering themselves, the victim, and the community. Students will be trained on properly transferring evidence and information collected at the traffi crash scene to reports and diagrams for courtroom testimony (**Table 3**) [4].

Advanced crash investigation

The prerequisite for this 2-week course is the completion of basic crash investigation. The course involves interpreting tire marks, road scars, advanced measuring methods, vehicle damage analysis, and vehicle behavior during a crash. If applicants have problems with math from basic crash investigation, they should practice algebra skills prior to signing up for the course, since there is a noticeable emphasis on math formulas that relate to collisions (**Table 4**) [4].

Traffic safety specialist (TSS) certification component

The traffi safety specialist certifi ation was fi st established in Maryland to support the Maryland Strategic Highway Safety Plan [6]. Th s designation has been designed to recognize police offic s who have attained notable experience, education, training, and proficie cy in highway safety and traffi enforcement methods and procedures. The program began in 2011 and was opened to police agencies that perform law enforcement duties. In New Jersey, the designation was adopted by a criteria selection committee, in coordination with the Police Traffic Officers' Association and the state police.

The TSS program requires documentation of experience levels, training, job performance, and skill proficiency as a traffic officer. The first phase enables officers to be recognized for specialized training beyond the academy. *Level I* requirements include 3 years of independent patrol/traffic experience, speed detection device certification, standard field sobriety certification, and 30 points of earned electives.

One elective point is also awarded for each year (up to three points) of documented military experience, while up to three points (1 = Associates, 2 = Bachelors, 3 = Masters) are awarded for documented college experience. Officers may also obtain up to seven points for traffic safety awards received during their employment.

After the *TSS program-Level I* has been successfully approved, *TSS-Level II* requires applicants to have participated in local traffi studies or traffi safety implementation projects. They must also have 5 years of independent patrol/traffi experience; meet the *TSS-Level I* requirements; and be recognized as a crash reconstruction investigator, drug recognition expert, or commercial vehicle safety

Module	Content
1.1. Introduction to crash investigation	Overview of crash investigation techniques used to support increased prosecutable crash cases
2.1. Crash investigation background	Review of crash investigation techniques that include at scene traffic - forcement, prevention programs, and resuming traffic w patterns
3.1. Crash information from people	Identifi ation of conditions affecting the ability of drivers to operate a motor vehicle and pedestrian misconception of driver abilities and motor vehicle capabilities that lead to crashes. Examination of physiological and psychological factors affecting drivers and pedestrians and occupant kinematics concepts
4.1. Recognizing roadway evidence	Recognition, investigation, and recording of physical marks from plotting vehicle positions before, during, and after the impact at the crash scene
5.1 Crash scene measurements	Skills development includes gathering accurate measurements and recording important data from the crash scene. Roadway configur tions and intersections are examined to enhance student abilities on developing scale diagrams, while customized traffic emplates are utilized to identify grades and superelevations. Recognizing and recording physical roadway marks also assist investigators in plotting the vehicle position before, during, and after the impact
5.2 Crash scene measurements (traffic emplate)	Examples are presented on the limitation of measurements, along with techniques for using templates for diagramming crash scenes for and courtroom presentations
5.3 Crash scene measurements (fi ld sketches)	Further development and refi ement of skills to gather and record accurate measurements and data from the crash scene. Several roadway configur - tions and intersections are used to enhance student abilities on developing scale diagrams. Illustrations and drawings are also produced from photographs and other reference materials
6.1. Documenting crash scene evidence	Background information is presented on effectively documenting evidence with photography. Emphasis is placed on taking suffici t photographs, based on camera location at the crash scene for reconstruction and courtroom presentations
7.1Documenting crash scene evidence from vehicles	Vehicle damage assists in determining the cause of a crash. Matching vehicle damage to roadway marks and identifying the difference between contact and induced damage. Also, data collection priorities at the crash scene are featured
7.2 Tire inspection and documentation	Background information on tires and at-scene methodologies are examined relative to documenting evidence. Tire abnormalities, hydroplaning action, and other conditions are further examined
7.3 Evidence from lamps	Descriptions of lamp types and parts and research related to legal statutes. There is also a discussion on incandescent lamps and general terminology used for gaining a better understanding of hot and cold shock. Also, lamp inspection is addressed, as it relates to the crash scene

8.1 Basic mathematics review	Brief review of concepts and function of calculation tools enables students to successfully complete basic mathematical functions
8.2 Coeffici t of friction and drag factor	Drag sled and test skid procedures are demonstrated for determining grade and acceleration/deceleration. Emphasis is placed on the importance of locating, recognizing, measuring, and recording physical marks on the roadway at the crash scene
8.3 Estimating vehicle speed	Importance of locating, recognizing, measuring, and recording physical roadway marks and evidence is addressed, along with the procedure used for determining the coeffici t of friction and speed estimates for skid marks
8.4 Time and distance	Speed estimates are conducted to calculate time and distance factors of a crash

Table 3.
Basic crash investigation course content.

Module	Content
1.1. Introduction to advanced crash Investigation	Advanced crash investigation requires students to have knowledge of interpreting tire marks and road scars. Advanced measuring methods, vehicle damage analysis, and vehicle behavior during a crash are addressed in this course
2.1 Vehicle damage	Skills are presented for obtaining residual crash photographs and measurement of vehicle damage. The Cartesian coordinate system will be described along with the linear perimeter measurement system. Vehicle behavior in a crash, PDOF, measurement of residual damage, and determination of width or length of damage are also reviewed in this module
2.2 Vehicle damage review	Field exercise is conducted for obtaining residual damage photographs and vehicle damage measurements
3.1. Vehicle behavior in crashes	Newton's laws of motion [5] are related to vehicle behavior during a crash. Relationships between motion, acceleration, and action/reaction are applied to crash investigation and principle direction of force (PDOF)
4.1 Advanced math review	Information is presented for calculating speed estimates from simple skid and yaw marks, as well as other evidence found at the crash scene
4.2 Time-distance calculations	Relationship between time and distance in the crash sequence is reviewed and applied to calculate the vehicle location during a crash

4.3Work and energy	Work is done when a vehicle slides, while (kinetic) energy dissipates into heat energy, a tool for determining vehicle speed and a major part of crash investigation. Examples are presented on the amount of change produced or work done as there is change in velocity of an object
4.4Geometry and trigonometry review	Examples are presented on the use of angles, triangles, and trigonometry in crash investigation
4.5 Momentum	Collinear momentum analysis for pre-crash determination is developed
4.6 Two-dimensional momentum	Two-dimensional momentum project is conducted using "X" and "Y" axis information
4.7 Radius and critical speed yaw	Radius and critical speed yaw are presented and measured, in order to calculate an actual average vehicle speed from the marks
4.8 Airborne crashes	Speed of a vehicle will be calculated when it has left the road surface
4.9 Introduction to vectors	Effective formulas and methods are examined for completing vector diagrams, linear-momentum calculations, and scalar exercises

Table 4.
Advanced crash investigation course content.

inspector. A traffi study must be submitted by applicants, along with a proof of 30 earned elective points, in order to be eligible for equivalent college credits and the TSS credential that is recognized for court testimony (**Table 5**) [6].

Over 80 courses are approved in 9 designated areas at county-based police academies. Many national programs (i.e., FBI, IPTM, and NIMS) apply because they are held in New Jersey and often sponsored through NJDHTS and NJSP. One half point is awarded to courses that are 4 hours in duration, while maximum credit of 10 points will be received for 80-hour training programs. Quarterly meetings are held to review the awards with Kean University taking the leadership as committee chairperson for the 10-member committee. The program was piloted in February 2016 and continues to attract participants who are dedicated to promoting crash reduction and injury prevention initiatives.

A linear approach to traffi safety education has been used for student engagement as passengers, pedestrians, or bicyclists. Elementary students are trained with

Categories	**Level I**	**Level II**
Patrol experience	3 years	5 years
TSS level designation	N/A	Level I
Certifi ations	• Speed detection device • Field sobriety test cert.	• Reconstructionist, DRE, ACTAR cert., or traffic gineering cert., • Instructor cert.

Elective options* • Military (3 points) • Post-secondary (3 points) • Traffic wards (7 points)	30 points	30 points
Written reports	N/A	Traffic tudy written report—Participant level

Table 5.
Traffic safety specialist designation criteria.

awareness techniques and practice of injury prevention behaviors, while middle school students address traffi safety issues through assessment and remediation of unsafe conditions. High school students are fully engaged in preparing for their roles as drivers and/or safe passengers, but not without the help of parents and the community. Conversely, the enforcement component provides law enforcement professionals with analytical tools for identifying potential crash conditions and potential crash hazards, which may/or may not be apparent to the general public. Furthermore, office participation in the traffi safety specialist program enables them to share personal traffic afety accomplishments with their peers.

Effectiveness of the prevention model

Education component

Two studies were conducted to address the effectiveness of the education component. First, a 1-year pilot study was conducted in 2011 that involved teachers from the second largest school district in the state. Evaluations involved ease of usefulness and classroom benefit. A comparison of group differences was based on participation in monthly and weekly presentations. A t-test and ANOVA application was used for evaluation purposes. Next, a 3-year study was conducted with parents attending the Share the Keys, a parent orientation. It was speculated that the school-wide K-12 pilot program would bring community attention to traffi safety and further reduce crash rates over time and parents will actively remain involved in monitoring their novice teen's driving patterns.

K-12 traffic safety program

In fall 2011, the second largest school district in New Jersey participated in a pilot study of the traffi safety component. A coordinator training session and four auditorium programs were conducted by task force representatives. One hundred forty-two teachers used the 76 lesson plans and resources to instruct over 3600 students. Th s sample group represented elementary teachers (86%), high school teachers (8%), and kindergarten/middle school teachers (6%). The most frequently used lessons were on bicycle safety (27%) followed by pedestrian safety (22%). All

Lessonsplans	(N = 142)
Bicycle	27
Bus safety	8
Driver education	10
Pedestrian	22
Trafficsafety	17
Unspecifi d	16

Table 6.
Percentage of lesson plans reviewed.

high school teachers (10%) used the driver education unit because it corresponded with the New Jersey driver education manual (**Table 6**).

When the year ended, a brief survey on the usefulness of the lesson plans and resources was distributed to participating teachers. Self-reported responses (n = 137) addressed time appropriation for classroom use, age appropriateness of content, best features of lesson plans/resources, areas of improvement, and other comments. Seventy percent of the teachers reported that lessons were completed within a 45-minute class period. All respondents felt that the content was age appropriate for their grade level. When asked about the best features of the component, most mentioned content/activities (N = 53), followed by available lesson plans/cods (N = 39), and lessons were easily implemented (N = 25). Also, teachers confi med that these "interactive" resources were successfully used by substitute teachers.

Overall mean scores for usefulness of components and resources were established and then used to test the null hypothesis of no difference in user satisfaction of resources between monthly and weekly instructor groups. Participating teachers were asked to rate the usefulness of units/lesson plans and corresponding resources for their grade levels, based on the following scores: 1 = very useful, 2 = somewhat useful, and 3 = not useful. Overall responses ranged between 1.35 (introduction) and 1.54 (PowerPoints) with lesson plans and handouts receiving the highest rating of a 1.37 score. The overall, monthly, and weekly user group means for the lesson plan components and resources appear in **Table 7**.

Next, the t-test, assuming unequal variances are p = <0.05, df = 92, was conducted to determine differences between combined means of monthly and weekly user groups. The t-statistic (3.505) supported the failure to reject the null hypoth- esis; therefore, an ANOVA, two-factor without replication, was conducted using

	Overall mean (N = 142)	Monthlymean (N=93)	Weeklymean (N=39)
Introduction	1.35	1.39	1.31
Benchmark	1.49	1.64	1.35
Plans	1.37	1.52	1.23
Handouts	1.37	1.51	1.23
Parent info	1.42	1.47	1.37
Power points	1.54	1.8	1.29
Teacherresources	1.45	1.56	1.34

Table 7.
Usefulness ratings of lesson plan components and resources.

Source of variation	SS	Df	MS	F	P-value	Fcrit
Rows	0.061943	6	0.010324	0.999539	0.500216	4.283866
Columns	0.223779	1	0.223779	21.66598	0.003487	5.987378
Error	0.061971	6	0.010329			
Total	0.347693	13				

Table 8.
ANOVA results.

the mean for each category, in order to determine if there were differences between and within groups on usefulness averages for monthly and weekly instruction.

There appeared to be some variation between the usefulness scores for lesson plan components and resources.

However, the values differed between monthly and weekly users, who had rated the lesson plan/resources as being more useful than the other group (p = 0.003487). Th s fi ding indicates that a Type 1 error had occurred and a difference existed on usefulness of the traffic afety component (**Table 8**).

While further research is needed on difference in benefits for elementary, middle, and high school teachers, this study was successful in obtaining self-reported benefits and program enhancements. Forty percent of the respondents indicated that program changes were not needed and another group (N = 24) expressed the desire to continue the program in their district. Others felt strong about providing additional time for the lessons (N = 14), incorporating more movement/physical education/walking trips/demonstrations (N = 20) into the program, and additional videos (N = 6) would improve learning. Finally, 10 teachers reported that the lesson plans should be used in turnkey (before/after school) programs.

STK parent orientation program 3-year analysis

Fifty-three school districts participated in the 3-year study that netted 2817 parent/teen teams from 14 counties with a population of over 6,300,000 residents. Nearly half of the participants were from the South (45%), followed by North (42%) and Central (13%) regions of the state. The 2010 US Census county data was also used to identify demographics of this cohort. As the program expanded, the ethnic composition maintained a greater representation of Caucasians (55%) and fewer Asians (5%). Gender was evenly matched between males and females, and only 34% compared to 44% of the population at large had completed college and earned an income of $73,653.

The NHTSA Fatality Analysis Reporting System (FARS) was accessed to identify the fatality rates of individuals between the ages of 16 and 20 on New Jersey roadways. The updated GDL restrictions had played a role in helping to reduce roadway fatalities for this population with a 63% decline in fatal crashes of 16–20-year-old drivers. The fatality rate of teen drivers had steadily declined from 52 to 33 roadway deaths in New Jersey, during this period. *Share the Keys* remained a voluntary program and was adopted by 10% of the school districts to support parents in enforcing and restricting the driving activities of their teens.

Chi-square and independent-sample t-test analyses were conducted to determine the differences between baseline and follow-up responses in all *domains*: *subjective norms, parental behaviors*, and *reported teen driving behaviors*. Most notably, there was a 5% overall reduction in *follow-up* responses (15% vs. 20%) between the two periods. Levene's test for equality of variance was used to identify whether or not variance of scores for the two groups was the same. Percentages, means, standard deviations, and t-scores were calculated for each of the 19 behaviors being measured. No signifi ant differences between the means were present in the *subjective norms* and *reported teen driving domains*, while differences were found in the *parent behavior* and *practice driving hours domains* (**Table 9**).

Domains & categories	**Baseline (n = 2817)***			**Follow-up (n = 437)**			
	N(%)	M	SD	N(%)	M	SD	T
Subjective norms							
Pstyle 1—permissive	1315 (46)	1.53	0.49	190(43)	1.57	0.49	−1.21
Pstyle 2—uninvolved	125(05)	1.90	0.29	52 (11)	1.82	0.38	4.14
Pstyle3—authoritative	1237(44)	1.47	0.49	140 (33)	1.61	0.49	−5.32
Pstyle4—authoritarian	76 (03)	1.91	0.29	31 (07)	1.90	0.31	1.10

No response	64 (02)			24 (06)			
Self-reported good driving role model	2523 (90)	1.10	.31	407 (93)	1.07	.25	2.02
No response	294 (10)			30 (07)			
Parent behavior							
Understand GDL	2330 (87)*	1.13	0.33	366 (84)	1.16	0.36	−1.90**
No response	338 (13)			71 (16)			
Practice driving	2127 (80)*	1.20	0.40	345 (79)	1.21	0.40	−0.37
No response	541 (20)			92 (21)			
Enforce GDL	2264 (85)*	1.15	0.36	346 (79)	1.21	0.40	−2.75**
No response	404 (15)			91 (21)			
Control keys	2200 (83)*	1.18	0.38	285 (65)	1.35	0.47	−7.19
No response	468 (17)			152 (35)			
Practice driving hours							
0–3 hours	747 (27)	1.73	0.44	149 (34)	1.66	0.47	3.13**
3–5 hours	433 (15)	1.85	0.36	69 (15)	1.84	0.36	0.22
5–7 hours	460 (16)	1.84	0.37	52 (12)	1.88	0.32	−2.60**
7+ hours	908 (33)	1.68	0.47	120 (28)	1.73	0.44	−2.06**
No response	269 (09)			47 (11)			
Reported teen driving behavior							
Curfew	2489 (88)	1.12	0.32	344 (79)	1.21	0.41	−4.70
No response	328 (12)	1.19	0.40	93 (21)	1.16	0.37	1.70
Passenger	2268 (81)			366 (84)			
No response	549 (19)			71 (16)			
Seatbelt	2586 (92)	1.08	0.27	356 (82)	1.19	0.40	−5.41
No response	231 (08)			81 (08)			
Ask permission	2382 (85)	1.15	0.36	291 (67)	1.33	0.47	−7.61
No response	435 (15)			146 (33)			
No text/cell	2434 (86)	1.14	0.34	290 (67)	1.34	0.47	−8.51
No response	383 (14)			147 (33)			
No alcohol	2645 (94)	1.06	0.24	381 (87)	1.13	0.34	−4.03
No response	172 (06)			56 (13)			

$p > 0.05$ for comparison characteristics of the sample.
**$n = 2668$ sample size for the parent behavior domain found in post-survey*
***Sig. (two-tailed) value is equal to or less than 0.05.*

Table 9.
3-year analysis of group differences: Percentages, unadjusted means, and t-test results.

The t-test results confi med that there were no signifi ant changes between the baseline and follow-up responses for parenting behavior; therefore, the null hypothesis was accepted. Next, chi-square results were examined to further identify trends in behavior over time. Interestingly, a shift occurred in the *subjective norms domain* between *authoritative* (44–33%), *uninvolved* (5–11%), and *authoritarian parenting styles* (3–7%). *Parent behaviors* remained consistent in all categories except for *control keys* (83–65%) which decreased by 16%, 1 year later. There were few changes in the *practice driving hours domain*, while results were not as positive in the *reported teen driving behaviors domain*. Compliance with *passenger limits* (81–84%) increased, but all other driving behaviors had decreased between 7% and 19% over time and netted a 10% average reduction in teen compliance with safe driving behavior.

Additional independent MANOVA tests were conducted to examine the relationship of driving phases to *control keys* behaviors that had shifted over time in the t-test. Results confi med that fewer parents *control keys* (M_1 = 1.44, SD = 0.50; M_2 = 1.24, SD_2 = 0.43) during the probationary phase and afterward. Also, teen compliance with the *driving curfew* (M_1 = 1.41, SD_1 = 0.49; M_2 = 1.12, SD_2 = 0.32) and *passenger restrictions* (M_1 = 1.27, SD_1 = 0.44; M_2 = 1.09, SD_2 = 2.9) had decreased at the probationary phase, not when teens were fully licensed. *Ask permission* to drive appeared to be influenced by the licensing phase, since mean scores continually decreased over time (M_1 1.43, SD_1 = 0.50; M_2 = 1.31, SD_2 = 0.46; M_3 = 1.24, *SD* = 0.43).

The overall results identifi d that there were insignifi ant differences between baseline and follow-up scores in the 19 behaviors used to prove the null hypothesis of GDL compliance over time. However, the STK interactive intervention was also examined in relation to changes in *subjective norms* (i.e., parenting styles), during the three driving phases. The MANOVA results confi med changes in levels of parental involvement had occurred during the probationary phase of licensure, not at full licensure, as projected. *Authoritarian* roles increased during the probationary phase, while *authoritative* roles appeared to be strongest after teen drivers earned their basic license. *Permissive* and *uninvolved* roles remained relatively consistent throughout the three driving phases. Interestingly, all types of parents appeared to have benefited from attending the orientation, and they remained actively involved in enforcing the GDL restrictions, 1 year later.

In the *parent behavior domain*, parents reported comparable levels of involvement in the following behaviors: *understand GDL*, *practice driving*, and *enforce GDL*. *Control keys* of their teen drivers, the only category not mandated under the GDL, had considerably decreased over time. There also appeared to be a relationship between parental enforcement and teen driver compliance with the GDL restrictions, especially with *control keys* and *ask permission* to use the car. Although these behaviors netted a 65% compliance level, 36% of teens had received their "unrestricted" basic license 1 year later.

Th *practice driving hours domain* remained consistent, except during year 1 when a downward trend appeared in the 7+ hours behavior. There was potential signifi ance in several categories (Year 1–7+ hours, Year 3-0–3 hours, 5–7 hours, and 7+ hours), but proved to be insignifi ant based on the t-test results. As mentioned, authoritative parents tended to increase practice driving hours when teens received their basic license.

Enforcement component

Crash investigation program

Examination scores were used to assess whether or not students were successfully learning new information. Basic crash investigation test results were examined in **Table 10** to measure improvements over time. The pre–/posttest scores for 2015

	Mean	**Std. div.**	**Std. error mean**	**Lower**	**Upper**	**t**	**df**	**Sig. (two-tailed)**

Pair 1	**2015 pre−/post**	**20.8000**	**16.08864**	**5.087**	**9.29088**	**32.30912**	**4.088**	**9**	**0.003**
Pair 2	2016 pre−/post	34.70000	21.51511	6.80368	19.30902	50.09098	5.100	9	0.001
Pair 3	2017 pre−/post	34.0000	23.98147	7.58361	16.84469	51.15531	4.482	9	0.002

Table 10.
Paired sample incorrect test scores (95% confidence interval).

showed a mean reduction that ranged between 21 and 35% of "incorrect" answers. which realized a 50% increase over time (3/6%). Further examination of the baseline (pre-test) and posttest results confi med that overall reduction of incorrect answers had dropped between 12% (*tire marks)* and 50% (*reaction).* During 2016, in only one area (*understanding of imprints*), there was no difference between the pre-test and posttest scores. These results confi med instructional benefits that students receive from their crash investigation faculty. A paired sample t-test was conducted to evaluate the impact of preand posttest scores for basic crash investigation. There was a signifi ant decrease in the 2015 posttest scores (e.g., improvement) from Time 1 (M = 40.80050, SD = 20.08759) to Time 2 (M = 20.0000, SD = 7.78888, t (9) = 4.088, p < .003 (two-tailed)). The mean decrease in posttest scores was 20.8000 with a 95% confide ce interval ranging from 9.39088 to 32.30912. The eta-squared statistic (9.0) confi med this effect, too. There was an even greater variance between the 2016 pre-test and post- test scores from Time 1 (M = 55.50000, SD = 26.53405) to Time 2 (M = 20.8000, SD = 11.36075, t (9) = 5.1000, p < 0.001 (two-tailed)). The mean decrease in posttest scores was 34.70000 with a 95% confide ce interval ranging from 19.30902 to 50.09098. The eta-squared statistic (9.0) further confi med this effect in size. The variance was also signifi ant between the 2017 pre-test and posttest scores from Time 1 (M = 46.3000, SD = 23.54688) to Time 2 (M = 12.300, SD = 6.12917, t (9) = 4482, p < 0.001 (two-tailed)). The mean decrease in posttest scores was 34.0000 with a 95% confide ce interval ranging from 16.84969 to 51.15531. The etasquared statistic (9.0) further confi med this effect in size.

Traffic safety specialist program

The traffi safety specialist program is unique because it awards police offic s who have accomplished signifi ant experience, education, training, and proficie cy in highway safety and traffi law enforcement. Level 1 recognizes over 100 individual accomplishments, while Level 2 represents leaders of the traffi safety community. The fi st cohort of Level 2 designees recently received this award in October 2018. Offic s were required to document 3 years of independent traffi control experience, provide documentation of speed detection certifi ation, maintain the Standardized Field Sobriety Testing Certifi ation, and earn 30 professional credit points. Level 2 requirements are similar, except the office must submit a local traffi study for approval.

2018 Best practices: TSS-level 2

Since the program is relatively new, offic s have submitted traffi studies that address similar conditions. Two reviews involved intersections in North and South Jersey, while the third study addressed traffi safety around a school district. In addition to examining collected data, offic s sought legislative guidance on the proposed countermeasures. A brief review of each "best practice" appears in the remainder of this section.

South Jersey county intersection study

The purpose of this project was to study the intersection of a county road that needed an alternative method of traffi control to make an intersection safer for motorists. A thorough study of this intersection included photographing the intersection with different approach angles and distances, performance of two independent traffi counts during the morning and afternoon hours, and retrieving and analyzing crash data from 2013 through 2016, to examine potential factors such as pedestrian travel, roadway conditions, and obstructions of view. Crash reports from 2013 to 2016 were retrieved and examined for collisions within the intersection. The fi dings determined that there was a total of 17 collisions caused by vehicles from the northbound direction entering the travel lanes of the route.

Crash reports from 2013 to 2016 were examined for collisions within the intersection. The fi dings determined that there was a total of 17 collisions caused by vehicles from the northbound direction entering the travel lanes of the county route. There were four collisions as a result of vehicles traveling west and attempting to turn left onto a street and colliding with vehicles. The majority of collisions were caused by drivers traveling northbound and failed to yield or observe vehicle traffi along the county route. After reviewing the US Department of Transportation Manual on Uniformed Traffi Control Devices (MUTCD), it was established that a traffi signal within the intersection would not satisfy any of the nine warrants required to substantiate the creation of a traffic s nal.

Other alternatives were prohibiting left turns onto the route from vehicles traveling northbound. It was not feasible, as prohibiting left turns on northbound route would dramatically increase thoroughfare traffi and drivers would use the additional side streets for access. Prohibiting left turns from residential side streets would also force additional traffi to an intersection that was not wide enough to handle the increased vehicle fl w. Taking all information into consideration, erection of signage prior to the intersection showing "Intersection Ahead" and another showing "Cross Traffi Does Not Stop" prior to the intersection was deemed appropriate.

North Jersey County intersection study

A township traffi safety department received complaints from residents (e.g., one per month) about an intersection. One of the five crashes, occurring between January and February 2014, resulted in a vehicle winding up on the front lawn of a resident. While the section is straight and leveled with a minimum grade, it contains a curve and has a downhill east to west grade that levels out to the roadway. The speed limit is 25 mph and serves a through street.

Along with the installation of stop signs at the intersection, 100 feet of advanced warning markings were installed in the roadway in June 2002. Th s action was in response to a citizen's concern over increased intersection crashes. The goal was to warn both lanes of travel that vehicular traffi was mandated to stop at the approaching intersection. All traffi control devices are in compliance with the MUTCD. During this period, a brief study netted contributing crash factors that included driver inattention [7], failure to yield [8], and failure to obey traffi control [8]. Most were Right Angle Turns [2], while Left Turn and Side Swipes netted [1] each.

There are several recommendations that would improve visibility at this intersection. First, the light bulb needs to be changed to a LED on the 8" flashing signal. By using a LED light, the flashing light would be more visible during daylight hours when all of the crashes occurred. Th s action was taken and provides more advanced warning for motorists. Another recommendation to improve visibility is the solar panel LED-blinking stop signs. Driver inattention and failure to observe a traffi control device have contributed to crash circumstances, so more visible stop signs will add reinforcement to the drivers who do not come to a complete stop. The solar panel will avoid electrical costs to operate, and the battery will be charged during sunlight hours. The fi al recommendation is to install an "Intersection Ahead" (W-1) sign on the northbound and southbound roadways. A supplementary sign stating "Intersection Ahead" should be posted below the intersection symbol to alert drivers of the upcoming intersection.

North Jersey traffic stop at the school zone study

Safety concerns associated with all-way stops include pedestrians, bicyclists, and all road users expecting other road users to stop. In accordance with the MUTCD, the decision to install multi-way stop controls requires an engineering study to determine the following:

Where traffi control signals are justifi d, the multi-way stop is an interim measure that can be installed quickly to control traffic while arrangements are being made for the installation of that signal. Five or more reported crashes on a road, during a 12-month period, make it a candidate for a multi-way stop installation. Such crashes include right-turn and left- urn collisions as well as right-angle collisions:

- (1) The vehicular volume entering the intersection from the major street approaches (total of both approaches) averages at least 300 vehicles per hour for any 8 hours of an average day; and (2) the combined vehicular, pedestrian, and bicycle volume entering the intersection from the minor street approaches (total of both approaches) averages at least 200 units per hour for the same 8 hours, with an average delay to minor street vehicular traffi of at least 30 seconds per vehicle during the highest hour, but (3) if the 85th percentile approach speed of the major street traffic exceeds 40 mph, the minimum vehicular volume warrants are 70% of the values provided in items 1 and 2.
- Another criterion that may be considered in an engineering study is location where a road user, after stopping, cannot see confli ting traffi and is not able to negotiate the intersection unless confli ting cross traffic s also required to stop.
- It should also be noted that the high school is located at this intersection, which greatly increases pedestrian traffi

The proposed intersection fell below the vehicular volume to be considered an all-way stop, but the large volume of crashes would have been avoided with such a treatment. An all-way traffi stop intersection was created at the designated location and included two 30-inch retrorefl ctive stop signs (two already existed), two 10-foot u-channel posts and base posts, four all-way supplemental plaques, and restriping of the intersection to include four 4-inch hashed crosswalks, four stop bars, rumble strips at all four approaches, and yellow curbing 25 ft of all crosswalks and 50 ft of all stop signs. After implementation, zero motor vehicle crashes have been reported at the intersection.

Conclusion

An effective Strategic Prevention Framework (SPF) was established by SAMSHA and adopted in New Jersey to better understand community traffi safety needs and promote sustainability for achieving and maintaining long-term results. Once safety issues are identifi d through analysis of crash data, available resources are reserved such as manpower, equipment, or training to support a formal plan based on the logic model. Actions are then implemented to address the identifi d safety needs. A major focus of the education component has been teacher approval of traffi safety program, while little is known about student benefits related to participation. While the lessons were adaptable to a 45-minute period, they promote student engagement that will have a long-term effect on community safety. A second 3-year study, involving parents of novice drivers, will be conducted in summer 2019. Past research has proven that parental monitoring of teen's driving behaviors serves as a prevention of uncheck reckless behavior that is likely to occur without direct supervision.

Most importantly, community leadership needs to be promoting outreach activities based on this prevention model. Th ough the university, a student internship program has been established to bring the message to the community, which is similar to the role of TSS offic s. Interestingly, the crash investigation component continues to operate with law enforcement serving as traffi safety advocates and monitoring the environment on preventing potential crashes. Best practices have been established in New Jersey by the four TSS-Level 2 leaders, with over 100 more TSS offic s waiting to qualify for this second tier. Future plans involve an outreach program for offic s to develop traffi safety programs in their communities while qualifying for the TSS-Level 2 designation. Th s injury prevention framework has proven to be effective in utilizing education and enforcement to advocate for the adoption of traffi safety goals. Further research needs to be done, especially in the area of roadway injury reduction involving community support.

Acknowledgements

Technical support was received from Derek DIStaso (NJSP) and the New Jersey Division of Highway Safety on oversight and funding of law enforcement education. The contributions of Nicholas Schock (President) and the Police Traffi Offic s Association of New Jersey are nationally recognized as "best practices" for reducing traffic talities. Also, Peter Cokelet, Michael Peraset, and Michael Tullio serve as resource specialists and editors for the development and revision of technical resources. Lastly, students and police offic s are recognized for their continuing efforts regarding improvement of traffic afety in their communities..

Conflict of interest

C.M.K., S.P., R.A.J., A.P., and G.P. have no competing interests to disclose. C.M.K. has received a grant from the New Jersey Division of Highway Traffi Safety, the organization that funded this study. G.P. represents the funding organization, but is not directly affected by publication of this research study.

Author details

Claudia M. Knezek[1*], Susan Polirstok[1], Roxie James[2], Anthony Pittman3 and Gary Poedubicky[3,4]

College of Education, Kean University, NJ, USA

College of Natural and Applied Health Sciences, Kean University, NJ, USA

College of Education, Acting Dean of Education, Kean University, NJ, USA

Division of Highway Traffic fety, State of New Jersey, Trenton, NJ, USA

*Address all correspondence to: cknezek@kean.edu

References

[1] Anderson-Carpenter KD, Watson-Thompson J, Jones M. Using communities of practice to support implementation of evidence-based prevention strategies. Journal of Community Practice. 2014;**22**:176-188

[2] SAMHSA. Prevention Approaches. 2017. Available from: http://www.samhsa.gov/capt/ practicing-effective-prevention/ prevention-approaches

[3] Children's Hospital of Philadelphia. Driving Th ough the Eyes of Teens: A Research Report of the Children's Hospital of Philadelphia and State Farm; 2007. Available from: www.chop.edu/injury

[4] NJDHTS. Police Resources: Crash Investigation Training; 2018. Available from: http://www.nj.gov/oag/hts/index. html

[5] NJDHTS. State of New Jersey Strategic Highway Safety Plan. New Jersey Department of Transportation and New Jersey Division of Highway Traffic fety; 2016

[6] McGuire T. Maryland Traffi Safety Specialist Program Reference Manual, Version 8.1, 2016-2017. p. 9

[7] McCall, J. New Jersey Traffi Safety Learning Progression Component. NJDOE Cumulative Progress Indicators; 2009. Available from: http://www.state. nj.us/education/cccs/standards/2/2-1-D. htm

[8] Graham T. Road Safety Education in the Scottish Curriculum. Scottish Executive Central Research Unit; 2000. Available from: http://www.scotland. gov.uk/Publications/2000/05/1818ec65- 25c3-40fc-8ca8-91935d728b30

[9] Kean University. The New Jersey K-12 Traffi Safety Component. p. 2012. Unpublished

[10] School Drug Education and Road Awareness (SDERA). Principles for Road Safety Education, Australia Research Summary. 2009. Available from: http:// www.det.wa.edu.au/sdera/detcms/ navigation/road-safety/principles-for-school-road-safety-education/

[11] ATSSA. Putting Safety in the Safe Routes to School Program, A School Administrator's Guide. The American Traffi Safety Services Association Roadway Safety Outreach Program. 2009. Available from: https://docs.alsde.edu/ documents/92/SRTSAdminGuide.pdf

[12] NJDOE. New Jersey Student Learning Standards for Comprehensive Health and Physical Education. New Jersey Department of Education; 2015

7

Influence of Tribological Parameters on the Railway Wheel Derailment

George Tumanishvili, Tengiz Nadiradze and Giorgi Tumanishvili

Abstract

At present, Nadal's formula is used for prediction of derailment that contains a limited number of parameters. Besides, insuffici t study of laws of variation of the noted parameters and ignorance of the influence of other parameters on the derailment complicate solution of the problem. The sliding distance and the relative sliding velocity are the most sensitive factors contributing to the destruction of the third body. Moreover, increased friction coeffici t between the steering surfaces of the wheel and rail promotes climbing of a wheel on the rail and derailment. Dependences of the main parameters, influencing the destruction of the third body, the sliding distance and the relative sliding velocity on the rail track curvature, and difference of diameters of wheels of the wheelset and the non-roundness of one of the wheels of the wheelset are shown in the work. The methods for estimation of the third body destruction degree and consideration in Nadal's formula of the additional criterion of impossibility of the wheel rolling on the contact point of the wheel and rail steering surfaces, containing a value of this contact point advancing, which in turn depends on the angle of attack, are proposed.

Keywords: tribological parameters, derailment, sliding distance, wear rate, third body

Introduction

The correlation of lateral and vertical forces, angle of inclination of the wheel flange, and coeffici t of friction between the wheel flange and rail gauge are considered as main parameters acting on derailment [1]. Avoidance of derailment and ensuring durability of the wheelsets, rails, brake shoes, etc. are vital for railways for both safety and economic reasons [2]. The prevalent case of derailments is climbing of a wheel on the rail that is influenced by such main parameters as the flange angle, vertical and lateral forces, angle of attack, friction factors, etc. There are many works devoted to these phenomena [1, 3–13] that indicate urgency of the problem.

The climbing of the wheel on the rail is stipulated by the tribological, geometric, and dynamical parameters of the wheel-rail interaction. For the solution of the problem, qualitative and quantitative estimations of influences of these parameters are necessary.

The well-known Nadal's criterion (1896) of the wheel climb derailments uses the lateral-to-vertical force limit (L/V limit) of a single wheel [8] depending on the angle of inclination of the wheel flange and friction coefficie t. However, the latter changes in the wide range and laws of this variation are not suffici tly studied. Besides, the wheel climb derailments generally occur in situations where the climbing wheel experiences a high lateral force at great angle of attack, which is not considered in Nadal's formula. The number of experimental researches confi ms the insuffici t reliability of Nadal's criterion [14–16].

In **Figure 1** is shown a rail with a trace left on it after the wheel climbing [17]. The trace starts on the rail lateral surface and then passes on the rail tread surface.

The mechanism of generation and development of this trace is not studied suffici tly yet and needs additional researches [18]. Besides, according to this paper, friction coeffici t in the contact zone of the wheel and rail reaches 0.5 and more at derailments.

The wheel climbing on the rail is also promoted by decreasing the rail radius of curvature and deviation of the axle of symmetry of the wheelset from radial position (increased angle of attack) that causes advancement of the wheel flange and railhead lateral surface contact point.

As it is known, a vertical axis of symmetry of the rail is inclined by 20° according to the standard. Defl ction of the rail in the opposite direction that decreases the angle of inclination of the wheel flange is especially dangerous for the wheel climbing on the rail.

A creep is typical for the wheel and rail interaction. Different parts of interacting surfaces of the wheels and rails need to have different properties. Friction factor for the wheel flange and rail gauge face should be as low as possible—less than 0.1. Excessively high friction of the tread surfaces causes severe wear, plastic fl w, and fatigue, and low friction can cause poor traction and braking. For tread surfaces of the wheel and rail, friction factor should not be less than 0.25 and greater than 0.4. Optimal value of the fi tion factor for these surfaces is 0.35 [12].

At common operational conditions, interacting surfaces are covered by various types of boundary layers—products of interaction of the surfaces and the environment, friction modifie s, etc.—that prevent a direct contact of the rubbing surfaces. Depending on the friction conditions, properties of the environment and surfaces, these layers may have various tribological properties that will have a great influence on the boundary friction [19–23]. Th s is confi med by the results of the experimental researches in the inert gas environment and vacuum that excludes the possibility of interaction with the environment [2]. Under such conditions, unhindered seizure and intensive wear rate are observed.

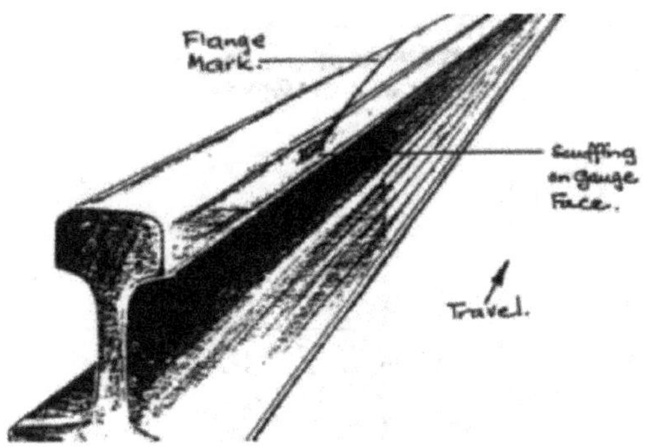

Figure 1.
The trace of the wheel climbing on the rail.

To prevent the aforementioned undesirable phenomena, it is important to provide the third body with due properties in the contact zone, control of the friction factor, and protection of the third body from destruction. However, until recently, despite considerable quantity of works, devoted to the study of dependences between wheel/rail and wheel/brake shoe friction forces and their durability, expected results are not obtained yet.

Our attention in the paper is mainly focused on the parameters that promote destruction of the third body. Some geometric features of the wheel and rail interaction and their influence on the friction path (sliding distance) and relative sliding velocity are shown. A corrected criterion of the wheel derailment is developed.

Dependence of the friction coefficient on the degree of destruction of the third body

The phenomenon of seizure is typical for interacting surfaces. Th s may occur when the third body is destructed and the surfaces are juvenile (free from dirty, oxide films and adsorbed layers) and are approached suffici tly. Seizure of the interacting surfaces leads to the most dangerous and dominating kind of deterioration—scuffing.

For prevention of this phenomenon, they try to improve the tribological characteristics of the contact zone (improve properties of contacting surfaces and their ambient by applying the friction modifie s), stabilize the boundary layers, minimize a sliding distance and relative sliding velocity, etc. As it is noted in [23], the variation of the friction coeffici t is mainly caused by changing a composition of the interfacial layer (the "third body") between interacting surfaces. Our experimental researches have shown that for the given friction modifie , the variation of the friction coeffici t mainly depends on the degree of destruction of the third body. An increase of the relative sliding velocity leads to an increase of the friction power and the contact temperature and decrease of the lubricant viscosity, film thickness, and friction force (friction coeffici t). It corresponds to the "negative friction" in **Figure 2**, where a friction/creep relationship is shown [24].

Worsening of the working conditions is caused by the partial, unit seizures and nonprogressive damage of the third body in the separate unit places (**Figure 3**) that corresponds to the separate small impulses of the friction moment. In **Figure 3** are shown the stages of damage of the interacting surfaces due to seizures and scuffing of the surfaces.

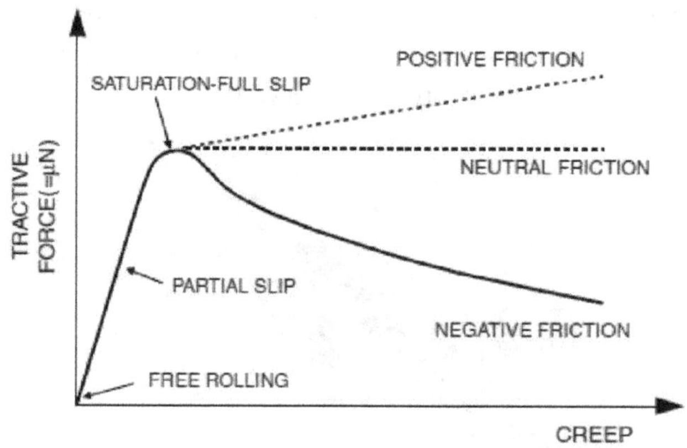

Figure 2.
Friction/creep relationship.

The further extension of destruction of the third body in the multiple places leads to the multiple damage of the third body, multiple adhesive junctions of micro-asperities, disruption of these junctions, and comparatively increased impulses of the friction moment and to "neutral friction."

A progression of the third body destruction leads to spacious, discontinuous third body, adhesive junctions of micro-asperities, disruption of these junctions, and increase of the friction forces ("positive friction"). As it is seen from **Figure 2**, negative, neutral, and positive behaviors of the friction forces are stipulated by the degree of destruction of the third (**Figure 3**) body (the unit, multiple, narrow strip, and whole area).

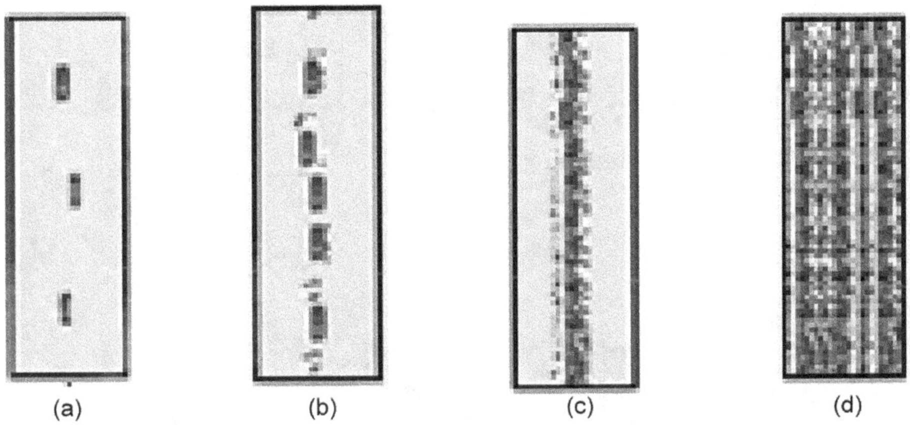

Figure 3.
The stages of damage of the interacting surfaces due to seizures and scuffing of the surfaces: (a) unit seizures, (b) multiple seizures, (c) seizures in the form of the narrow strip, and (d) seizures on the whole area of the roller.

Therefore, for ensuring the high wear resistance and stable friction force in the contact zone of the wheels and rails, it is necessary to provide continuous film of the third body with due properties between interacting surfaces. Consequently, a condition of destruction of the third body can be used as basics for estimation of the friction coeffici t and the damages for the given peculiarities of the surface materials.

The various dominant damage types, wear rate, and friction coeffici t are characteristic for various relative sliding. In **Figure 4** is shown dependence of the friction coeffici t on the relative sliding and expected kind of surface damage. Th ee zones can be distinguished in **Figure 4**. In zone 1 and at the beginning of zone 2, deformations of the subsurface layers reach the maximum values, and the interacting surfaces undergo cyclic deformations. With the rise of relative sliding, the contact temperature gradually increases, decreasing viscosity of the third body [24] and the friction factor that reaches the minimum value. At full separation of the interacting surfaces by the third body, the tribo-technical properties of the contact zone mainly depend on the properties of the third body, and they provide high wear resistance of the interacting surfaces and relatively stable friction coeffici t.

In zone 2 the separate small impulses of the friction moment and adhesive wear of low intensity correspond to destruction of the third body in the separate unit and multiple places, and balance between destruction and restoration of the third body is observed that stipulates the "mild" and "sever" wear [25]. In zone 3 destruction of the third body takes place in the narrow strips that passes then into whole area of interacting surfaces, resulting in rise of the friction coeffici t, its instability, wear rate (reaching "catastrophic" wear), and scuffing.

So, we have three stages of variation of the friction coeffici t and wear: at continuous third body, at reversible discontinuous third body, and at irreversible discontinuous third body. The fi st stage is characterized by the minimal wear rate and stable friction factor. The second stage is characterized by the small constant and variable components of the friction coeffici t. In terms of tribological characteristics, stages 1 and 2 indicate the acceptable working conditions of the tribological system. In contrast to this, stage 3 is characterized by the sharp increase of the constant and variable components of the friction coeffici t, wear rate ("catastrophic wear"), vibrations, and noise, and operation in this zone is not admissible.

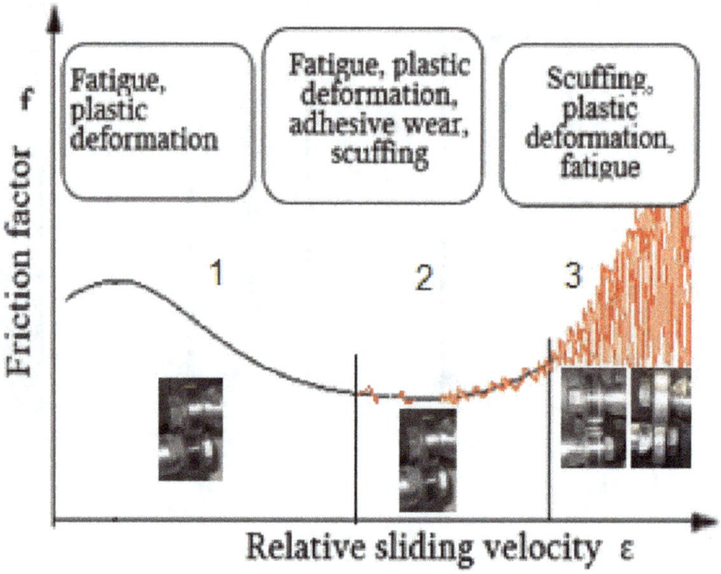

Figure 4.
Dependence of the coefficient of friction (f) on the relative sliding (ε) and expected kind of surface damage.

The friction coeffici t is minimum and stable in the fi st and second zones, and its value depends mainly on the rheological properties of the third body. In the third zone, the friction coeffici t is sharply increased and instable, and its value depends on the working conditions and properties of the surfaces, friction modifie s, and environment. The signs of the beginning of the third body are instability of the friction (coeffici t) moment, vibrations, and noise, and at visual observation in the laboratory conditions, the signs of scuffing are noticeable. Its prediction is possible with the use of the tables and graphs considering the given friction modifie , working conditions and environment properties, as well as the criterion of destruction of the third body [26].

Some geometrical peculiarities of the wheel and rail interaction

The geometrical features of the wheel and rail interaction are stipulated by the designs of the rail track/bogie, wheel/rail, and their technical state. At lateral displacement of the wheelset relative to the rail, a contact point from the tread surfaces passes on the wheel flange root and rail corner, and the wheel and rail tread surfaces separate from each other. At further lateral displacement of the wheelset, the contact point passes on their steering surfaces, and the angle of inclination of the wheel flange increases up to 70°.

It is difficult to predict and control the friction forces, wear rate of various types, vibrations, and noise of the heavy loaded interacting surfaces of the railway transport running gear that decreases traffi safety, increases energy loses on friction, etc. Many works are devoted to the researches of dependences of the tribological properties on various factors [1, 3–6], though their mechanisms of generation and variation are not always entirely clear that complicates the revelation of parameters influencing them [7, 8].

There are many reasons of generation of vibrations and noise at movement of the train, part of which are well studied and predictable, and ways of their decrease are known. The interacting surfaces of the wheels and rails are characterized by the various types of irregularities, 5–20 mm gaps in the rail joints, where the rail tread surfaces are spaced by 0.5–2 mm in the vertical direction; the various wear traces (rail corrugation, fatigue, etc.) and deviations from the wheel roundness are the sources of vibrations and noise.

The wheel and rail interaction is accompanied by the forced and self-excited vibrations of various frequencies, as the main reason of the forced vibrations is considered macro-and micro-asperities of the rail (periodic and separate asperities) [9–13]. However, the main source of the self-vibrations is friction between the wheel and rail. It must be noted that to various working conditions of the heavy loaded contacting surfaces and wear types correspond typical micro-asperities, which can be different from the initial micro-asperities [27, 28]. The researches have shown an important role of the tread and steering surfaces in generation of the vibrations (self-vibrations) and noise, whose reasons are not studied suffici tly. There is quite vague information on the reasons of the self-vibrations generated at interaction of the wheel and rail [9].

Generation of vibrations of the heavy loaded interacting elements of the railway transport running gear is stipulated by the complex processes proceeding in the contact zone. As a result of interaction of the surfaces with the environment, they are coated by the layers of various physical and chemical origins that are the components of the third body in the contact zone and have a great influence on the tribological properties of the contacting surfaces. According to observations by Godet, dry friction is largely determined not by the properties of materials of the contacting pair but by the characteristics of the structure and composition of the thin film that is formed on the surfaces of both bodies because of compaction of the wear product and its chemical composition and oxidation. Destination of the third body in the tribological systems is separation of the contacting surfaces, providing with the stable friction forces of proper values and protection of the surfaces against damage of various types. Tribological properties of the third body greatly depend on the initial properties of its component elements and features of the contact zone. The sliding velocity, power and thermal loading, and the sliding distance have especially great influence on the destruction of the third body. For providing the stability of the third body in the contact zone of the wheels and rails and reduction of the derailment probability, energy consumed on traction, environment pollution, and maintenance expenses, the decrease of the sliding distance and relative sliding is especially important.

The wheel/rail squeal in curves is the most common type of vibrations and noise. It is especially typical for high-speed movements, when because of various reasons, the relative sliding and sliding distance increase. Th s contributes destruction of the third body, seizure of the surfaces at direct contact, subsequent destruction of the seized surfaces, and instability of the friction forces and relative movement of surfaces.

Many negative phenomena (wear, noise, vibrations) are generated because of the wheel sliding on the rail. For elimination of the wheel sliding in the curves, the wheel tread surface is given a conical form with the intention of making the outer wheel to roll on the greater diameter passing the greater distance than the inner wheel and rotate both wheels through the equal angles, maintaining this way radial position of the wheelset axle. However, this intention can be realized only for a certain combination of such parameters, as radius of the rail track curvature, mass and speed of the rolling stock, friction coeffici t between the wheel and rail, etc. Therefore, practically the outer wheel rolls on the less diameter than necessary, and in the case of the free wheelset (without bogie), it falls behind the inner wheel, inclining the wheelset axle from the radial position.

In the case of the non-free wheelset, the bogie makes the wheelset maintain a radial position, forcing the outer wheel to roll the greater distance not to fall behind the inner wheel. Thereat, the outer wheel rotates through the greater angle than the inner one, and the wheelset axle is twisted. The angle of twist of the wheelset can increase up to the value that is stipulated by the friction force between the wheel and rail. When this angle of twist reaches the limited value, the wheel slides on the rail due to action of the wheelset axle elastic moment tending to bring it back to the equilibrium position.

Similarly, the wheel will slide on the rail at rolling in the straight rail track of the wheelset with the wheels of different diameters or with one wheel having an elliptical form. The mechanisms of the wheel sliding on the rail for the three noted cases are considered and explained in the next paragraphs.

Movement of the wagon wheelset in the curve

At pure rolling of the free wheelset (without bogie) in the curved rail track with radius of curvature R of the internal rail, its axle will be inclined from radial position because both wheels will have passed equal distances l. However, in the wagon wheelset rolling with velocity V, the outer wheel is constraint to maintain the radial position and pass greater distance $l + \Delta l$, rotating relative to the inner wheel in the clockwise direction if it is seen from axial direction A (**Figure 5**). At that, the wheelset axle is twisted through angle φ equal to the ratio of the difference Δl of the outer and inner arcs to the radius D/2 of the wheel tread surface, supposing that both wheels are rolling on the tread surfaces of equal diameters:

$$\varphi = 2\Delta l = D \quad (1)$$

From the drawing $\alpha = l/R = (l + \Delta l)/(R + \Delta R) = \Delta l/\Delta R$, from where

$$\Delta l = l\Delta R = R, \quad (2)$$

and therefore

$$\varphi = 2\, l\Delta R = DR. \quad (3)$$

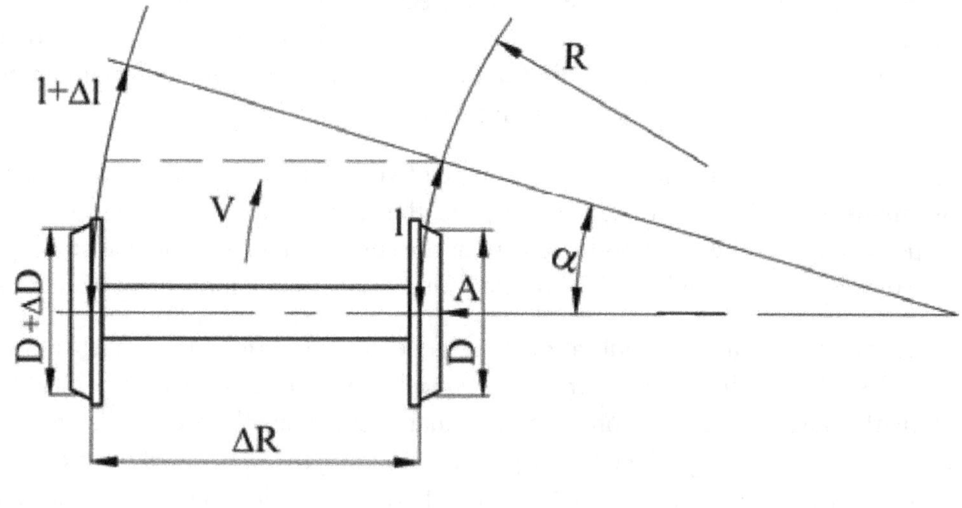

Figure 5.
Movement of the wagon wheelset in the curve and wheelset shaft slope from the radial position.

On the other hand, the maximum angle of twist of the wheelset axle φmax depends on the friction force

$$F = fQ \quad (4)$$

and is calculated by the known, from the resistance of materials, formula

$$\varphi_{max} = ML/I_p G, \quad (5)$$

where M is a torque caused by the friction force

$$M = FD/2 = fQD/2; \qquad (6)$$

f, friction coeffici t; Q, vertical load (half of the load on the wheelset) of the wheel on the rail; L, length of the wheelset axle; I_p, polar moment of inertia of the wheelset axle cross section; and G, modulus of rigidity (share modulus) of the axle material.

We determine distance between the worn-out segments of the rail or path l (at traveling this path, the wheels are rolling on the rail without sliding), at rolling of which the axle is twisted on the maximum angle φmax, from (3) replacing φ by φmax

$$l = DR\,\varphi_{max}/2\Delta R = MLDR{=}2I_pG\Delta R \qquad (7)$$

and putting the found l into (2) we obtain difference of the paths passed by the outer and inner wheels at which the axle is twisted on the maximum angle φmax

$$\Delta l = MLD/2I_pG. \qquad (8)$$

Movement of the wagon wheelset with the wheels of different diameters in the straight rail track

At rolling of the free wheelset (without bogie) with the wheels of different diameters D and D + ΔD in the straight rail track the distance l, the greater wheel passes a greater distance l + Δl, defl cting the wheelset axle from its perpendicular position relative to the rail track (**Figure 2a**). But in the wagon wheelset the axle being constraint to retain perpendicular position, the smaller wheel is forced to pass the same distance l + Δl and rotate relative to the greater wheel in the clockwise direction, if it is seen from axial direction A. At that, the wheelset axle is twisted through angle φ that is determined by formula (1), from where, considering (5), we obtain the value of Δl (see formula (8)) corresponding to the maximum angle of twist φ_{max}.

The following proportion can be written from the drawing: (l + Δl)/l = (D + ΔD)/D or Δl/l = ΔD/D, from which we obtain distance l between the worn-out segments at passing of which the wheelset axle will be twisted through angle φmax:

$$l = \Delta lD{=}\Delta D = MLD^2/2I_pG\Delta D \qquad (9)$$

Movement of the wagon wheelset with one elliptical wheel in the straight rail track

Consider a free wheelset with one wheel of diameter D and other elliptical wheel with the small D and bigger D + ΔD diameters moving in the straight rail track (**Figure 6a** and **b**).

At one revolution, these wheels will pass the different distances, correspondingly l and l + Δl, defl cting the wheelset axle from its perpendicular position relative to the rail track (**Figure 2a**). However, in the wagon wheelset the axle being constraint to retain perpendicular position, the wheel with diameter D is forced to pass the same (greater) distance l + Δl and rotate relative to the elliptical wheel in the clockwise direction if it is seen from axial direction A. At that, the wheelset axle is twisted through angle φ that is determined by formula (1).

The difference of distances passed by the wheels at one revolution is Δl = L−πD, where the length of the elliptical tread surface

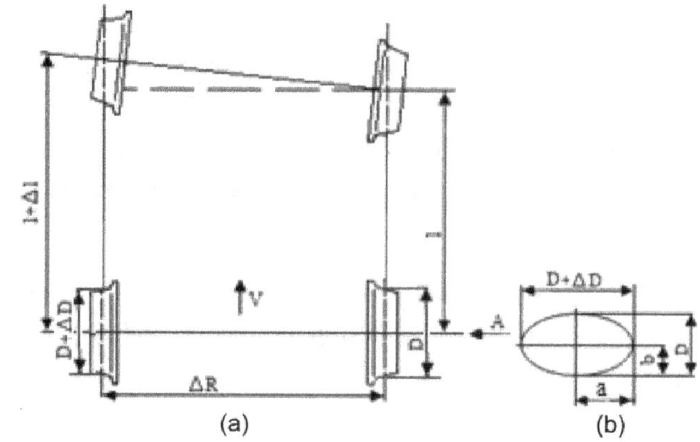

Figure 6.
Movement of the free wheelset in the straight rail track: (a) with the wheels of different diameters or with one elliptical wheel; (b) parameters of ellipticity.

$$L = \pi[3(a + b) - (3a + b)(a + 3b)] \tag{10}$$

or

$$\Delta l = \pi[3(a + b) - (3a + b)(a + 3b)] - \pi D \tag{11}$$

The value Δl^{I} corresponding to maximum angle of twist φmax is obtained con- sidering formula (5)

$$\Delta l^{I} = \varphi_{max} D/2 = MLD/2I_{p}G \tag{12}$$

The distance l at passing of which the wheelset axle will be twisted on the angle

φmax will be then

$$l = \pi D \Delta l^{I}/\Delta \tag{13}$$

In all the three cases considered above, at removing or decrease of the torque M acting on the wheel that takes place at its vertical vibrations when the friction force F decreases, the angle of twist of the axle will start to decrease. Suppose φmax falls down to zero during time t. Th s will take place at rotation of the inner wheel in the clockwise direction relative to the outer wheel on the angle φmax since the flange of the outer wheel is pressed on the rail and the friction force arisen between the flange and rail additionally restricts its movement. Obviously, during this time t the inner wheel will roll and slide simultaneously on the rail and the rolling and sliding distance on the rail will be

$$Sr = Vt \tag{14}$$

We note that the rolling and sliding distance on the wheel tread surface is

$$S_{w} = \Delta l + S_{r}. \tag{14'}$$

or for the variant of the elliptical wheel

$$S_{w} = \Delta l^{I} + S_{r} \tag{15}$$

here Δl or ΔlI is a sliding friction path and the wavelength of the worn-out rail (**Figure 3**)

$$W = las_{ers} \tag{16}$$

Th s value of the wavelength assumes that at release of the inner wheel, the friction force acting on it from the rail is zero. When the friction force differs from zero, the wavelength will be less since its both components will decrease and its value depends on the friction force magnitude.

To determine time t, we present the wheelset as a one-mass torsional vibratory system (**Figure 7a**), where C is a torsional rigidity of the wheelset axle and I, total moment of inertia of the inner wheel. Then, angle of twist φmax will fall down to zero in conformity with a law of free vibrations of this vibratory system during the period P/4 (**Figure 7b**).

At that, period of free vibrations

$$P = 2\pi\sqrt{I/C} \tag{17}$$

and consequently, time t will be

$$t = P/4 = \frac{\pi}{2}\sqrt{I/C} \tag{18}$$

The average velocity of the wheel contact point relative to the wheel center (Figure 8)

$$Vw = -\frac{D\varphi max}{2t} + Vr \tag{19}$$

where Vr = - V is a velocity of the rail contact point relative to the wheel center.

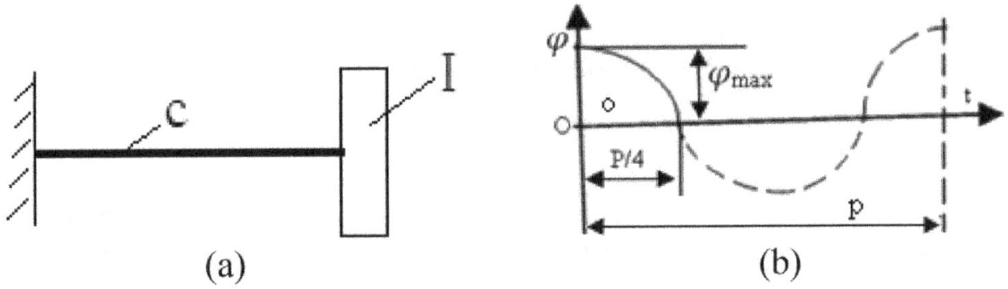

Figure 7.
(a) One-mass torsional vibratory system; (b) graph of the system free vibrations.

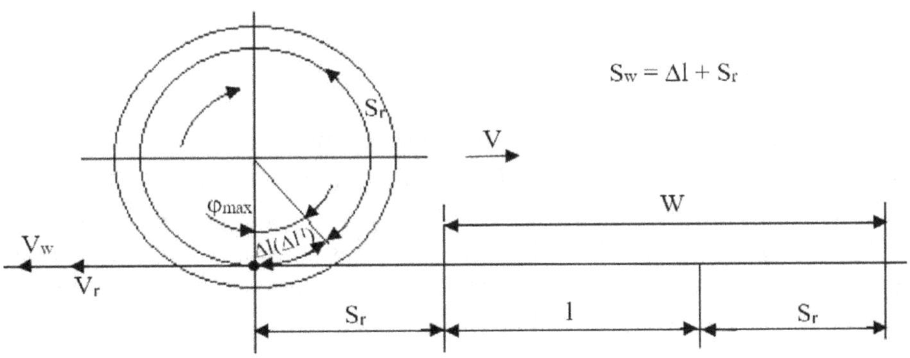

Figure 8.
The rolling and sliding distances on the rail and wheel.

We note that maximum velocity of the wheel contact point relative to the wheel center.

$$V_w^I = -\frac{A\omega D}{2} + \mathrm{Vr} = -\varphi\mathrm{max}\sqrt{\frac{C}{I}} \times \frac{\mathrm{D}}{2} + \mathrm{Vr} \tag{20}$$

where A = φmax is an amplitude of the wheelset shaft torsion vibrations and $\omega = \sqrt{C/I}$ is cyclic frequency of vibrations.

Sliding velocity

$$\mathrm{Vsl} = \mathrm{Vw} - \mathrm{Vr} \tag{21}$$

Relative sliding velocities

$$\mathrm{Kr} = \frac{V_{sl}}{V_r} \times 100\% \text{ and } \mathrm{Kw} = \frac{V_{sl}}{V_w} \times 100\% \tag{22}$$

The depth of the worn-out layer a year of the rail segment Sr.

$$\mathrm{h} = \mathrm{i}\Delta\mathrm{lN} \tag{23}$$

where i is the wear intensity and N, number of cycles which is determined as follows:

$$\mathrm{N} = \mathrm{N1N2N3N4} \tag{24}$$

where N1 is a number of the trains passing by a day; N2, number of wagons in the train; N3, number of wheels on one side of the wagon; and N4, number of days a year.

Conditions of derailment

Possibility of derailment or the wheel's rolling up on the rail is estimated by the criterion of the wheel flange contact point (point A, **Figure 9**) slipping down the rail lateral surface, based on the condition of equilibrium of forces acting on this point [24]. Lateral L and vertical V forces determined from the condition of equilibrium of these forces are.where N is a normal force; FI = fIN, friction force between the wheel flange and rail lateral surface; fI, friction coeffici t between these surfaces; and β, angle of inclination of the wheel flange.

$$\mathrm{L} = -\ \mathrm{N} \sin\beta - \mathrm{FI} \cos\beta \tag{25}$$

$$\mathrm{V} = \mathrm{N} \cos\beta + \mathrm{FI} \sin\beta \tag{26}$$

It should be noted that the forces acting on point A are interdependent and equalities (25) and (26) are only valid for limited values of forces L and V, since the rise of the friction force FI is limited by the friction coefficient fI. Therefore, at a certain ratio of forces L and V, the friction force FI can no longer balance the contact point A, which will slip down on the rail lateral surface, and it is considered on this ground that the wheel cannot roll up on the rail. At that, equalities (25) and become inequalities from where a criterion of impossibility of the wheel rolling up on the rail or derailment is obtained [24]:

$$\frac{L}{V} \leq \frac{\tan\beta - f^I}{1 + f^I \tan\beta} \tag{27}$$

However, at sign of equality (=) in (27) and to a certain extent at sign of inequality (<) also, the wheel can rotate about contact point A and roll up on the rail if such possibility exists or if moment of the force P acting on the wheel axle exceeds the moment of the vertical force V about contact point A (**Figure 10**). In other words, under such condition, two-point (O, A) contact of the wheel passes into one-point contact at A. In the fi st case (at sign =), the wheel will roll on the immobile point A with pure rolling, and in the second case (at sign <), the wheel will roll on the mobile point A creeping slowly down the rail lateral surface with combined rolling and sliding. Both cases lead to the wheel climbing the rail and derailment.

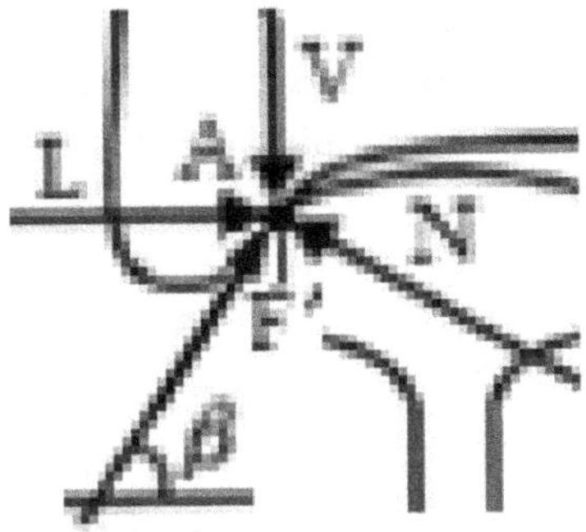

Figure 9.
Forces acting on the contact point a.

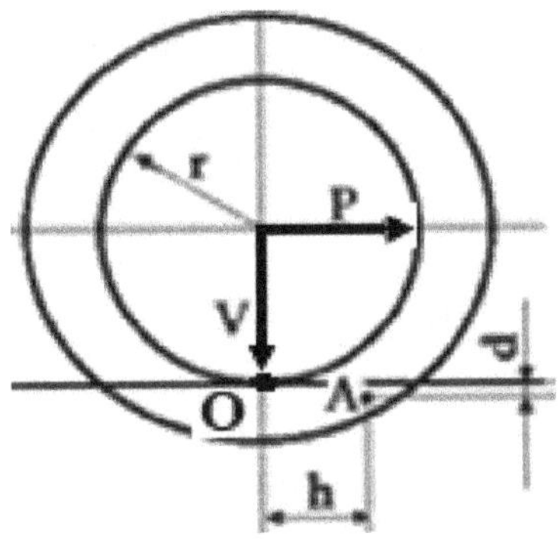

Figure 10.
Forces acting on the wheel axle.

Therefore, it is necessary to provide the criterion (27) with additional condition of impossibility of the wheel rolling on the contact point A, which, on the base of **Figure 10**, can be written as

$$Vh \geq P(r + d) \tag{28}$$

where h is the value of the climbing advance; r is the radius of the wheel rolling circle; d is the vertical coordinate of the contact point A.

Force P acting on the wheel axle cannot exceed the sum of the friction forces between the wheel and rail tread surfaces and between the wheel flange and rail lateral surface:

$$P \leq F + FI = fV + fIN \tag{29}$$

where f and fI are friction coeffici ts between the wheel and rail tread surfaces and the wheel flange and rail lateral surfaces correspondingly.

Determining N and V correspondingly from (25) and (26), substituting them into (29) and then putting obtained P into (28), from the latter we obtain the following criterion of impossibility of the derailment:

$$\frac{L}{V} \leq \frac{h\left(\sin\beta - f^{I}\cos\beta\right)}{(r+d)\left(f\cos\beta + ff^{I}\sin\beta + f^{I}\right)} \tag{30}$$

If this criterion is not satisfi d, the wheel starts to roll on the contact point A, and the contact between the wheel and rail tread surfaces is lost, or two-point contact at O and A passes into one-point contact at A. For obtaining a criterion of impossibility of the wheel rolling on the contact point A, it is necessary to put f = 0 in (30), which gives.

$$\frac{L}{V} \leq \frac{h\left(\sin\beta - f^{I}\cos\beta\right)}{(r+d)f^{I}} \tag{31}$$

The criteria (30) and (31) provide both, the wheel flange contact point sliding down the rail lateral surface and impossibility of the wheel rolling on this point. Besides, the criterion (30) ensures less value (more conservative) of the allowable ratio of the lateral and vertical forces L/V than criterion (27), while criterion (31), depending on the value of the climbing advance h, gives the ratio L/V less or more than criterion (27). For illustration, consider two variants of numerical data of the parameters:

a. β = 60o; f = 0.4; fI =0.1; h = 62 mm and r + d = 485 mm;

b. β = 60o; f = 0.4; fI =0.1; h = 88 mm and r + d = 482 mm.

Allowable maximum ratios L/V for these variants calculated by the criteria (27), (30), and (31) are given in the following table:

Variant	Criterion (27)	Criterion (30)	Criterion (31)
a	1.39	0.31	1.04
b	1.39	0.44	1.47

For analysis of the obtained results, suppose that ratio L/V = 1.3, i.e., criterion is satisfi d and derailment is not possible. However, it is seen from the table that for variant (a) neither criteria (30) nor (31) are satisfi d and both predict derailment. For variant (b), criterion (30) is not satisfi d, or it predicts derailment, and criterion (31) is satisfi d, i.e., by this criterion, derailment is not possible. Th s means that the wheel starts to roll on the contact point A and two-point (O, A) contact passes into one-point contact at A. Then, this contact point slides down the rail lateral surface, the two-point contact restores, and so on, this process is repeated. However, at passing from two-point (O, A) contact into one-point contact at A, the lateral and vertical forces on the steering surfaces increase. Typical for these surfaces, increased relative sliding increases the power and thermal loads in the contact of these surfaces, generating the convenient conditions for destruction of the third body. Th s results in sharp increase of the cohesion forces, scuffing, and friction coeffici t that promotes climbing of the wheel flange on the rail lateral surface. Th s is confi med by the numerous laboratory researches carried out by us as well as the trace of the wheel climbing on the railhead lateral surface (**Figure 1**) that has a form of scuffing.

Thus, it is expedient to estimate possibility of derailment by criterion (30), since it provides both, the wheel flange contact point sliding down the rail lateral surface and impossibility of the wheel rolling on the same point, and ensures less value (more conservative) of the allowable ratio of the lateral and vertical forces L/V than criteria (31) and (27).

Conclusion remarks and future works

Prediction and avoiding of derailment are the most important problems of which many scientific works are devoted for their solution but the desirable results are not obtained yet. The survey of the literature and our experience show that the derailment is especially influenced by the friction coeffici t that is not predictable, and in contrast to other parameters, it varies in a wide range.

It is shown that for prediction of the friction coeffici t and providing its stability, it is necessary to provide the contact zone with the continuous and restorable third body of due properties.

The main results of the paper can be formed as follows:

- A friction factor as well as other tribological properties of interacting surfaces depends on the properties and degree of destruction of the third body.
- The sharp increase of the friction factor in the contact zone of steering surfaces indicates a beginning of the irreversible (progressive) destruction of the third body that contributes to the wheel climbing on the rail.
- For avoidance of derailment, decreasing the wear rate and ensuring suffici t durability of the rails, wheelsets, and brake shoes, a continuous or reversible third body must be provided in the contact zone.
- Destruction of the third body in the laboratory conditions is proposed to determine by the flash of the friction moment or criterion of destruction of the third body.
- A criterion of impossibility of derailment providing additionally impossibility of the wheel rolling on the wheel flange contact point is offered, which ensures less value (more conservative) of the allowable ratio of lateral and vertical forces than Nadal's formula.

For solution of the problem of derailment, an experimental–theoretical approach is needed because of the lack of comprehensive theoretical model of the wheel climbing on the rail.

Due to the existence of materials with quite different designations and properties in the contact zone, many new unanswered problems rise. They are related with the further increase of the derailment criterion informativity and precision, providing the contact zone with the third body having due properties, conditions of formation, and destruction of the third body. They also concern to the tribological properties of the interacting metal and nonmetal surfaces, direct interaction of their juvenile surfaces and generation of the strong adhesion bonds, cold welding, destruction and wear of the surfaces, variation of the value, and instability of the friction coeffici t.

On the base of experimental researches, we have ascertained dependence of the friction coeffici t on the degree of destruction of the third body for the conditions of various relative sliding velocities, speeds, materials of interacting surfaces, roughness of the surfaces, friction modifie s, and loads at which the range of variation of the acting parameters is quite wide and therefore continuation of researches is needed.

Author details

George Tumanishvili*, Tengiz Nadiradze and Giorgi Tumanishvili Institute of Machine Mechanics, Tbilisi, Georgia

*Address all correspondence to: ge.tumanishvili@gmail.com

References

[1] Iijima H, Yoshida H, Suzuki K, Yasuda Y. Special edition paper. A Study on the Prevention of Wheel-Climb Derailment at Low Speed Ranges. JR EAST Technical Review-No. 30

[2] Harris CL, Wyn-Roberts D. Friction and wear studies in ultra-high vacuum and the evaluation of electrical slip rings. In: Proceedings of the Institution of Mechanical Engineers, Conference Proceedings. Vol. 183; 1968. pp. 50-60

[3] Huimin W, Nicholas W. Railway Vehicle Derailment and Prevention, Chapter 8 in Handbook of Railway Vehicle Dynamics, Edited by Simon Iwnicki. Taylor & Francis Group, LLC; 2006. pp. 209-239

[4] Shust WC, Elkins JA, Kalay S, El- Sibaie M, Wheel-Climb Derailment Tests using AAR Track Loading Vehicle, Association of American Railroads Report R-910; 1997

[5] Huimin W, Xinggao S, Nicholas W. Flange Climb Derailment Criteria and Wheel/Rail Profile Management and Maintenance Guidelines for Transit Operations. Transportation Technology Center, Inc. (TTCI). Washington, D.C: Pueblo, CO.; 2005

[6] Weinstock H. Wheel Climb Derailment Criteria for Evaluation of Rail Vehicle Safety, Proceedings of ASME Winter Annual Meeting, 84-WA/ RT-1. LA: New Orleans; 1984

[7] Thompson DJ, Honk-Steel AD, Jones CJC, ISVR, Allen PD, Hsu SS, Iwnicki SD, MMU. Project A3 – Railway noise: curve squeal, roughness growth, friction and wear; 20th June, 2003

[8] Wu H, Elkins J, Investigation of Wheel Flange Climb Derailment Criteria, Association of American Railroads Report R-931; July 1999

[9] Cristol-Bulthe AL, Desplanques Y, Degallaix G. Coupling between friction physical mechanisms and transient thermal phenomena involved in pad– disc contact during railway braking. Laboratoire de Mrecanique de Lille (CNRS UMR 8107), Ecole Centrale de Lille, BP 48, F-59651 Villeneuve d'Ascq Cedex, France. Available online 23 May 2007

[10] Elkins J, Wu H, New Criteria for Flange Climb Derailment, IEEE/ASME Joint Railroad. Conference, Newark, NJ, USA; April 4–6, 2000. pp. 1-7

[11] Rhee SK, Jacko MG, Tsang PHS. The role of friction film in friction, wear and noise of automotive brakes. Wear. 1991; 146(1):89-97

[12] Vasic G, Franklin FJ, Kapoor A. Prepared for the Railway Safety and Standards board. University of Sheffield. Report: RRUK/A2/1; 2003

[13] Weinstock H. Wheel climb derailment criteria for evaluation of rail vehicle safety. In: Proceedings of the ASME winter annual meeting; 1984. pp. 1–7

[14] Iwnicki S. Handbook of Railway Vehicle Dynamics2. © 2006 by Taylor & Francis Group, LLC. p. 38

[15] Braghin F, Bruni S, Diana G. Experimental and numerical investigation on the derailment of a railway wheelset with solid axle. Vehicle System Dynamics. 2006;44:305-325

[16] Clementson J, Evans J. The use of dynamic simulation in the investigation of derailment incidents. Vehicle System Dynamics. 2002;37:338-349

[17] Engineering Manual Track and Rolling Stock. TMC 213 Derailment Investigation - Track and Rolling Stock. Version 1.0; August 2011

[18] Brian Marquis, Robert Greif. Application of Nadal Limit in the Prediction of Wheel Climb Derailment. In: Proceedings of the ASME/ASCE/ IEEE 2011 Joint Rail Conference. JRC2011. March 16–18, 2011. Pueblo, Colorado, USA

[19] Ahmed NS, Nassar AM. Lubrication and lubricants. In: Tribol. -Fundam. Adv. 2013. pp. 55-76

[20] Yifei M, Turner KT, Szlufarska I. Friction laws at the nanoscale. Nature. 2009;457:26. DOI: 10.1038/07748

[21] Hou K, Kalousek J, Magel E. Rheological model of solid layer in rolling contact. Wear. 1997;211:134-140

[22] Arthur A, Jannik T. Design of a Test rig for Railway Curve Squealing Noise. Sweden: Chalmers University of Technology Gothenburg; 2017

[23] Drozdov YN, Pavlov VG, Puchkov VN. Friction and Wear in the Extreme Conditions (in Russian). Mashinostroenie: Moscow; 1986. p. 224

[24] Eadie DT, Kalousek J, Chiddik KC. The role of high positive friction (HPF) modifier in the control of short pitch corrugations and related phenomena. Wear. 2002;253:185-192

[25] Lewis R, Dwyer-Joyce RS. Wear mechanisms and transitions in railway wheel steels. Proceedings of the institution of mechanical engineers, part Journal of Engineering Tribology. 2004;218(6):467-478

[26] Tumanishvili G, Natriashvili T, Nadiradze T. Perfection of technical characteristics of the railway transport system Europe-Caucasus-Asia (TRACECA) in the book. In: Sladkowski A, editor. Transport Systems and Delivery of Cargo on East-West Routes, Springer. 2018. pp. 303-368

[27] Elkins JA, Wu H. Angle of attack and distance-based criteria for flange climb derailment. Vehicle System Dynamics. 1999;33:291-305

[28] Ishida H, Miyamoto T, Maebashi E, Doi H, Iida K, Furukawa A. Safety assessment for flange climb derailment of trains running at low speeds on sharp curves. Quarterly Report of RTRI (Japan). 2006;47:65-71

8

Approaches for Modeling User's Acceptance of Innovative Transportation Technologies and Systems

Stefano de Luca, Roberta Di Pace and Facundo Storani

Abstract

The gradual penetration of new transport modes and/or new technologies (advanced information systems, automotive technologies, etc.) requires effective theoretical paradigms able to interpret and model transportation system users' propensity to purchase and use them. Along with the traditional approaches mainly based on random utility theory, it is a common opinion that numerous nonquantitative variables (such as psychological factors, attitudes, perceptions, etc.) may affect users' behaviors. Different traditional approaches and more advanced ones (e.g. hybrid choice model (HCM) with latent variables, theory of planned behaviour, regret theory, prospect theory, etc.) may be identifi d and properly applied in the literature. In particular, the chapter will focus on the hybrid choice modeling with latent variables, aiming to incorporate users' perceptions, attitudes and concerns in order to model the user's propensity to use and the willingness to buy a new technology. The methodology overview and the results of the application at real data are discussed.

Keywords: transportation technologies, transportation systems, intelligent transportation systems, user's acceptance, choice modeling

Introduction

The diffusion and market penetration of new technologies are becoming a crucial point for transport system analysts and decision-makers. The main issues are regarding a correct understanding of the phenomena and the simulation of different possible operational scenarios.

Among the several new technologies aiming to let the transportation system be more effici t and sustainable, two main issues continue to be challenging tasks:

- interpreting and modeling users' behaviour towards these new technologies and
- assessing the potential environmental impacts.

Both issues are highly correlated as, without an effective interpretation of users' behaviour, no reliable estimation of the market penetration, and/or the corresponding impacts could ever be obtained.

Within the cited context, the consumer choice theory based on the random utility theory (RUT) may be considered the more effective and practical approach to model and forecast user's behaviour, but it is a common opinion that consolidated random utility model (RUM) formulations may lead to neglect the numerous nonquantitative factors that may affect users' perceptions and behaviors. As a matter of fact, psychological factors, such as attitudes, concerns and perceptions, may play a signifi ant role which should be explicitly modeled. On the other hand, collecting psychological factors could be a time- and cost-consuming activity. Furthermore, real-world applications must rely on theoretical paradigms easily implementable in order to allow the estimation of users' choices in different technological scenarios.

As clearly stated in the current literature, the propensity to adopt a new technology, and, in particular, an alternative fuel vehicle (AFV), is mainly affected not only by instrumental attributes of the technology of interest (alternative) and of the competing technologies (other alternatives) and by personal attitudes (attitudes) of the consumer (user) not depending on the alternatives but also on the consumer personal feelings and on the socioeconomic context in which he/she lives. Several recent analyses have pointed out the necessity to take into account attributes considering the perceptions and the attitudes of the users. For example, see [1]. The main issues of the literature refer to:

- How to better collect information about users' attitudes, concerns and/or perceptions; an "ad-hoc" survey aiming to "collect" attitudes/perception needs to be designed.
- The assessment of the methodology to be pursued in accordance with the RUT [2–5], which is the richest and by far the most widely used theoretical paradigm for modeling transport-related choices and, more generally, with discrete choice modeling (i.e. models representing choices made among discrete alternatives).

Indeed, even though RUMs models usually adopted in demand modeling are suitable for the representation of the choice process, these are not applicable to represent the perceptions and attitudes [6]. The issue was addressed in the literature through the hybrid choice models (HCM) [7–11] based on attitude investigations trying to infer the role of psychological factors with latent variables within a discrete choice modeling framework.

In particular, several studies aiming to overstep the boundary of RUMs have been conducted by Ortúzar and Hutt [12] and by McFadden [13], which around the 1980s investigated the possibility to include subjective variables in a discrete choice modeling. Starting from the approach proposed by Jöreskog [14] focusing on the investigation between latent variables and the measurement of the perception indicators, several researchers contributed to the assessment of the methodological framework of hybrid choice models (e.g. see [15–17]).

The book chapter is organized as follows: Section 2 focuses on modeling overview, while Section 3 focuses on survey design; the quantitative preliminary analyses and the model specifi ation are, respectively, discussed in Sections 4 and 5. The corresponding results of two case studies are displayed and discussed in Sections 2 to 5.

Conceptual overview

The HCM based on RUT is a discrete choice model which integrates and simultaneously estimates different types of sub-models into a unique structure. If the HCM includes a latent variable model, it is possible to take into account the effects of users' latent attitudes, perceptions and concerns (i.e. integrated choice and latent variable (ICLV) model).

Adopting the standardized notation for path analysis, **Figure 1** introduces the general structure of an ICLV and allows to comprehend the different sub-models that defi e an ICLV: the latent variable model and the discrete choice model. In particular, the ellipses represent the unobservable (latent) variables, the rectangles represent the observable variables, and the circles represent the error variance or disturbance terms.

Since the latent variables (attitudes, perceptions and concerns) cannot be directly observed and measured from a revealed choice or a stated preference experiment [18], they have to be modeled and then indirectly identified starting from a set of indicators. The latent variable model allows to identify and measure these unobservable variables as a function of the indicators, in order to include them in a choice model.

Mathematically, a latent variable is treated as a random variable; the latent variable is specified through a *structural equation* formalizing it as a function of several parameters and a random error term. With regard to the relationship between indicators and latent variables, it can be formalized through a *measurement equation*, in which each observed psychological indicator is a function of a latent variable and a random error term. In general, each latent variable may be part of more than one measurement equation.

Finally, in accordance with the RUT, the utility U^i_j that an individual i associates with an alternative j is considered as a function of explanatory variables. The latent variables are included in the utility function of the alternatives as explanatory variables.

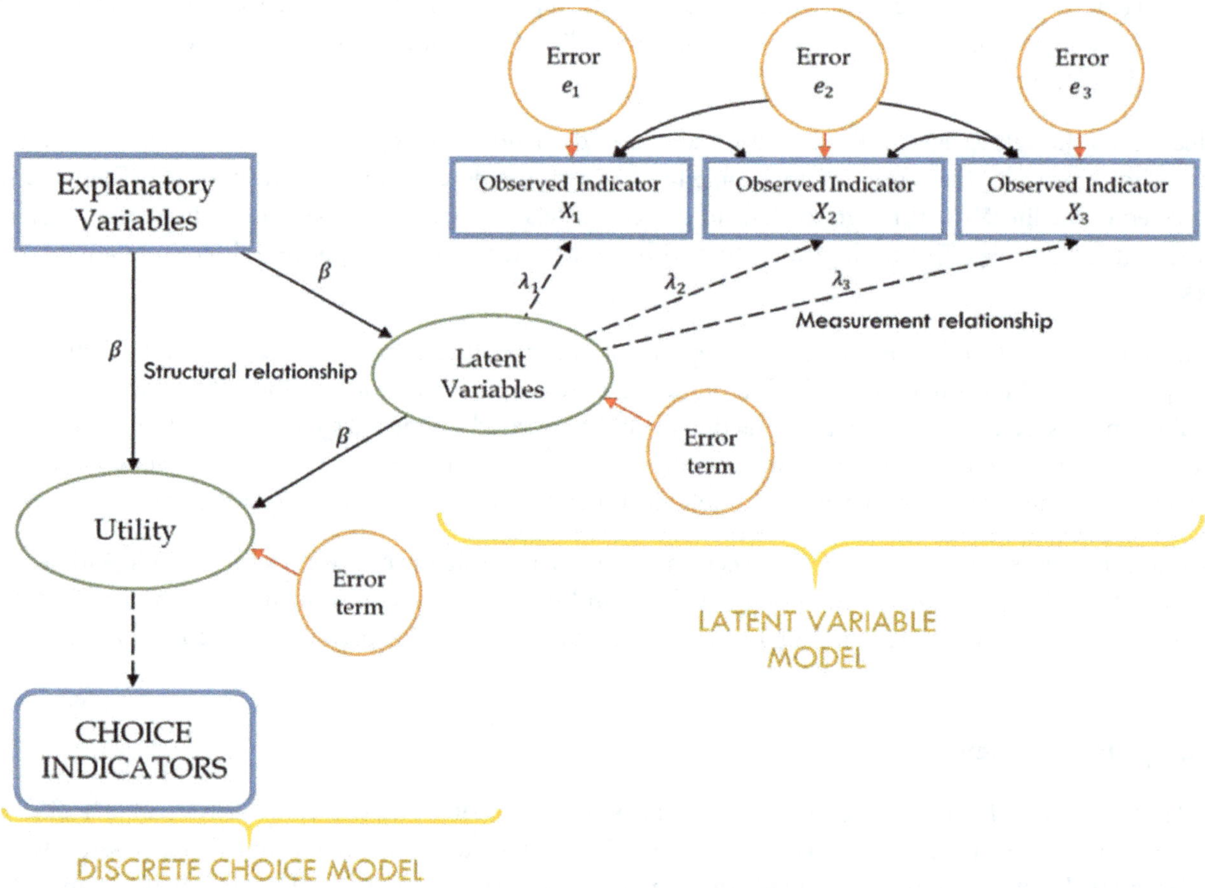

Figure 1.
Scheme of a hybrid choice model (HCM).

Survey design issues

One of the main issues in the specifi ation of an ICLV model consists of the observation and the measurement of the attitudes/concerns/perceptions. The survey design is a crucial part since it should allow to characterize the respondents, to come up with respondents' attitudes/concerns/perceptions and to measure them.

Usually, a survey is structured in different subsections aiming to collect various information from the respondents. First, (a) socioeconomic and (b) household characteristics have to be inquired; then specific sections aimed to capture users' attitudes, concerns and perceptions should be specifi ally designed, for instance, the users' attitudes and concerns that may influence the willingness to adopt/ use/purchase a new technology and (d) the users' perceptions with regard to the advantages and disadvantages of the technology under investigation. Finally, it is necessary to collect (e) the users' propensity to adopt/use/purchase a new technology. In this case, different scenarios (usually not real but realistic) should be carefully designed in order to cover the possible range of the involved decisional variables.

Even though the literature is consolidated on survey design with respect to Sections (a), (b) and (e), Sections (c) and (d) need to be specifi ally discussed.

As introduced before, one of the main issues related to the specifi ation and estimation of an HCM relies on how to collect users' attitudes (to observe and quantify them). Since attitudes are entities constructed to represent certain underlying response tendencies, they cannot be measured directly, but they could be inferred by studying behaviour which, in turn, might be reasonably assumed to indicate the attitudes themselves.

The behaviour may be one that occurs in a natural setting or in a simulated situation. In general, different approaches to measure attitudes may be pursued:

- *Direct observation*: this approach is based on either observing the actual behaviour of people or directly asking to state their feelings regarding the issue being studied. If the aim of the study requires the collection of information from a large number of individuals, this approach is not very practical. Moreover, the observation of peoples' behaviour may reveal the direction of the underlying attitude (i.e. whether it is positive or negative), but it cannot as easily indicate the magnitude or strength of the attitude itself, even when the behaviour is the outcome of the attitude being studied.
- *Direct questioning*: this approach consists of asking to a set of individuals what their feelings are, as a self-report technique. It serves only for a limited purpose of classifying respondents as *favorable*, *unfavorable* or *indifferent* towards a psychological object. An underlying issue with this technique is the fact that the individuals may possess certain attitudes and behave accordingly to them but may not be consciously aware of them, providing involuntary false statements to the direct questionings.

Th s method has two approaches to question the individuals: *direct* question on the investigated attitude (e.g. how important is the environment) and *indirect* question (e.g. do you normally buy...).

In general, direct questioning is the most pursued approach since it allows to control the investigated context and requires smaller times and costs. The method application requires the scale of measurement defin tion. Although the literature has proposed different scales as Thurstone, Likert and Bogardus, the Likert scale [19] has the most flex ble, robust and easy to implement scale of measurement.

Latent variables may be classifi d as attitudes [20, 21], perceptions [22] and concerns [23]:

- *Attitudes* refer to the users' characteristics and to their approach in real-life society and can be related to the alternatives (alternative-related attitudes) or not (non-alternative-related attitudes). They can be collected through direct or indirect questioning, but indirect questioning seems the most appropriate approach [24, 25].
- *Perceptions* are usually interpreted as alternative-related and refer to the users' interpretation and reaction to a stimulus [21]. They can be gathered through direct questioning only.
- *Concerns* may be related to a specific problem/issue. They may depend on the choice context (e.g. the concern towards the environment may depend on the specific problem/activity carried out). They can be collected through direct questioning only.

Within the aforementioned conceptual framework, it is suggested to design a direct questioning survey considering two different types of questions to be submitted to the respondents: direct and indirect questions. An overview of two examples of survey design is provided in the following.

Example 1: the HySolarKit case study

A fi st example refers to the HySolarKit [26] case study. The questionnaire described in this section was designed [24] to investigate the role of latent factors in the choice of a new automotive technology which aims to electrify/hybridize existing vehicles through an aftermarket kit which can be recharged by the grid but also by solar power (the HySolarKit). The experiment was applied to the case study of the Salerno municipality which is the capital city of Salerno province (region of Campania, southern Italy).

The fi st section of the questionnaire aimed to collect users' socioeconomic, activity-related attributes and household-vehicle characteristic information; therefore, respondents were provided with direct and indirect questions.

In particular, direct questions were about fuel consumption, vehicle reliability, vehicle design and the environmental impact; indirect questions were about three main latent behaviors: the fuel consumption, the vehicle design and the environmental impact. A detailed description is displayed in **Tables 1** and **2.**

The questionnaire was completed through a second section based on installation cost scenarios. In particular, each respondent was faced with two scenarios based on different installation costs (ranging from 500 to 4000 €).

Respondents were provided with a brief description of the technology and its main characteristics: how it works, how it is installed, the different performances (e.g. power, acceleration, speed), the environmental and fuel consumption benefits that can be achieved and the operating time. A brief overview is displayed in more detail in **Table 3** [27].

Example 2: the electric vehicle case study

The second example was about electric vehicle (EV) market penetration [28]. The questionnaire was designed with the aim to investigate the different attributes/ determinants that may influence the decision to purchase an electric vehicle.

Indicator	Statement
Q_{cons}	The vehicle fuel consumption signifi antly influences my choice in purchasing a new car
Q_{rel}	The vehicle reliability signifi antly influences my choice in purchasing a new car
Q_{design}	The vehicle design signifi antly influences my choice in purchasing a new car
Q_{env}	The evaluation of the environmental impact signifi antly influences my choice in purchasing a new car
Q_{price}	The price signifi antly influences my choice in purchasing a new car

Table 1.
Psychological statements—Direct questions.

Attitude towards fuel consumption

Indicator	Statement
$I_{cons}1$	The consumption and the energy class signifi antly influence my choice in purchasing an appliance
$I_{cons}2$	I am usually attentive to the special offers of electric operators
$I_{cons}3$	My home bulbs are energy effici t
$I_{cons}4$	I usually evaluate the car effici cy with respect to the car cost mileage
$I_{cons}5$	I normally compare the fuel prices among different petrol stations
$I_{cons}6$	When driving I am not willing to behave to reduce the environmental impacts (my driving behaviour is normally aggressive)
Attitude towards vehicle design	
Indicator	Statement
$I_{design}1$	When parking I am usually careful to avoid having my car damaged
$I_{design}2$	I often read journals of design
$I_{design}3$	When furnishing I am willing to buy pieces with modern design features and original details
$I_{design}4$	I am willing to go to the body shop mechanic not only for major damages
$I_{design}5$	I am willing to install not standard equipment (such as antitheft lock shaft) n my own car
Attitude towards environmental impacts	
Indicator	Statement
$I_{env}1$	I often control the exhaust/emission system of my car
$I_{env}2$	In consciously make the separate waste collection
$I_{env}3$	I really enjoy spending my free time in parks and green areas to breathe clean area
$I_{env}4$	How much do you agree with following sentence: We must act and make decisions to reduce emissions of greenhouse gases
$I_{env}5$	How much do you agree with following sentence: The government should invest in low energy impact
$I_{env}6$	I am not willing to use the car during weekend to protect the environment and then reduce air pollution

Table 2.
Psychological statements—Indirect questions.

The fi st section of the questionnaire aims to gather information about the users' socioeconomic characteristics, the characteristics of the owned household vehicles and the psychometric indicators of the latent variables. Particularly, two types of

Your owned car (internal combustion engine, ICEV)	**Your owned car with the HSK**

Brief description of the HSK

The hybridizing equipment (see figu e above) consists of:

- In-wheel electric engine
- Auxiliary lithium-ion battery pack
- Additional control system that interacts with existing vehicle components and optimizes energy fl ws
- Flexible photovoltaic panels installed on vehicle roof and hood

A mild parallel hybrid structure is obtained by substituting/integrating the rear wheels with in-wheel motors. In this way, the vehicle can operate in pure electric mode (when ICEV is switched off or disconnected by the front wheels) or in hybrid mode (when the ICEV drives the front wheels and the rear in-wheel motors operate in traction mode or in generation mode)

The battery can be recharged by:

- Rear wheels, when operating in generation mode
- Photovoltaic panels
- A regular electric power outlet, when the vehicle is connected to the grid power in plug-in mode

Technological features (averages)*		
	ICEV	ICEV + HSK
Power	P	P + (30%) x P
$Speed_{max}$	V	V or 40 km/h in pure electric mode
Acceleration (acceleration time from 0 to 100 km/h)	A	A-(25%) x A
Consumptions	C	C-(20%) x C
Emissions	E	E-(20%) x E
Operating time	T	T or 15 m in pure electric mode

Table 3.
Overview of the two alternatives in the choice context.

attitudes and two types of perceptions were inquired. The investigated attitudes were towards the environment and about the vehicle's technical features, while the investigated perceptions referred to the advantages and disadvantages of electric vehicles that may affect users' willingness to purchase them. To this end, several direct/ indirect questions were specifi ally designed adopting a five-point Likert scale (ranging from strongly disagree to strongly agree). The psychological statements used as indicators of those unobservable latent variables are detailed in **Table 4.**

The second section of the questionnaire contains the users' choice behaviour in buying a new car. To this end, a Renault Zoe as the electric alternative and a dieselfuelled Renault Clio are considered as conventional vehicles (CV).

Attitudes towards the environment	
Indicator	**Statement**
F_cons	The vehicle fuel consumption signifi antly influences my choice in purchasing a new car
F_poll	I care about the amount of pollution generated by a car when it's being used
Attitudes about technical features	
Indicator	Statement
F_tech_fea	The vehicle technical features signifi antly influence my choice in purchasing a new car
TF_power	I prefer driving a car with a powerful engine
TF_top_speed	When I am choosing a car, I fi d myself spending a lot of time checking out differences in terms of top speed (km/h)
TF_accel	The immediate acceleration increases my driving pleasure
TF_range	The vehicle range is very signifi ant with respect to my mobility needs in everyday life
Advantages	
Indicator	Statement
AD_{Red_CO2}	I am interested in EV to contribute to the emissions reduction
$AD_{Efficiency}$	Compared to a normal car, EV are superior in terms of energy effici cy
AD_{Red_poll}	I believe using EV can signifi antly reduce the acoustic pollution in cities
AD_{Less_parts}	I prefer having a car with less moving parts
Disadvantages	
Indicator	Statement
DIS_{infr}	When driving an EV, I would always be worried about the lack of charging points
DIS_{red_fea}	Compared to a normal car, EV are inferior in terms of performances
DIS_{batt_range}	The low range of the battery is a real disadvantage

Table 4.
Psychological statements of the survey.

These two alternatives were compared in terms of their technical features and then hypothesized that the interviewee has a budget enough to buy a new car to be used in urban areas. The monthly cost of buying and fuelling the conventional vehicle is calculated, considering 8-year monthly payments to buy it and an estimated fuel cost to drive it for 10,000 km per year. Th s results in a total cost of 192 € per month (**Table 5**).

After the comparison, each respondent was faced with five monthly cost scenario setup for the electric vehicle. They started from the same value calculated for the diesel vehicle (i.e. 192 €/month), and then it was considered that the EV would have cost 10, 20, 30 and 40% more (€211, €230, €250 and €270 per month, respectively).

Preliminary analyses

In the case of HCM, descriptive and statistical analyses of the collected answers regarding the perception indicators are necessary for two main reasons: fi stly, they allow supporting the soundness of the experimental design setup; secondly, they allow deriving fi st correlations between observed behaviour, collected attributes and investigated attitudes/perceptions/concerns. Therefore, results from preliminary analyses may give important insights on the survey robustness and a useful guidance for the model specifi ation. In general, the analyses may be divided in basic preliminary analyses based on the analysis of *Cronbach's alpha* which focuses on the evaluation of the internal consistency of the answers and the evaluation of *mean and standard deviations* and other advanced analysis, as the *factor analysis* which aims to identify the latent variables. A further and more detailed explanation of the methods is provided below.

Vehicle A—Electric Renault ZOE Life R90		**Vehicle B—Diesel Renault Clio 1.5 dCi Life**

Technical features (averages)*		
	Electric	*Gasoline*
Power	92 CV	75 CV
Top speed	135 km/h	168 km/h
Acceleration (0–100 km/h)	13.4 s	14.3 s
Consumption	10.9 km/kWh	30.3 km/l
CO_2 emissions	0 g/km	85 g/km
Range	240 km	1364 km

Table 5.
Comparison of the two alternatives in the choice context.

Cronbach's alpha

The Cronbach's alpha, α (or coeffici t alpha), is a measure of the internal consistency (or reliability) of the responses to multiple questions that are meant to measure a specific latent variable in a survey using a Likert scale. Th s indicator aims to tell whether the survey was accurately designed, and the questions were not answered randomly.

For each latent variable, it is necessary to have at least two indirect questions. The higher the number of questions, the better the latent variable would be measured. These questions, if possible, should be a mix of "+keyed" and "-keyed", depending if each one is positively correlated to the latent variable or negatively (**Table 6**).

When a survey intends to measure more than one latent variable, it is recommended to alternate their questions. Th s strategy combined to the mix of "+ keyed" and "- keyed" is employed in order to encourage respondents to be more aware of each item and the response provided and, therefore, increases the probability of gathering valid responding. If not, respondents may realize which latent

Scoring	Very inaccurate	Moderately inaccurate	Neither inaccurate nor accurate	Moderately accurate	Very accurate
+ keyed	1	2	3	4	5
- keyed	5	4	3	2	1

Table 6.
Scores for a five-point Likert scale.

variable is being measured, and they might tend to answer with the same response if the questions are all equally keyed. In this case, the total variance will be lower, and the relation with other variables in the study will be underestimated.

The Cronbach's alpha is calculated for each group of questions that measure a specific latent variable. Given $X = Y_1 + Y_2 + \dots + Y_K$ the sum of the scores of the K questions for each respondent, Cronbach's alpha is obtained as

$$\alpha = \frac{K}{K-1}\left(1 - \frac{\sum_{i=1}^{K}\sigma^2_{Y_i}}{\sigma^2_X}\right) \tag{1}$$

where:

K is the number of items (questions).

$^{\sigma 2}X$ is the variance of the observed total scores.

$^{\sigma 2}Y_i$ is the variance of the scores of each item i.

A rule commonly accepted to interpret the values of Cronbach's alpha when used with a Likert scale is (**Table 7**):

Mean and standard deviation

An easy quantitative/qualitative analysis to perform on the data collected consists of calculating the mean value and the standard deviation of the responses to the Likert scale for each question and then evaluating whether the responses are consistent (they have a low deviation) or if the mean value corresponds to the one expected.

Factor analysis

Given Y_i is an observable random variable, with mean μ_i, with $i = 1, \dots, p$ where p is the number of observed variables (the number of indirect questions), the factor analysis is a statistical method to investigate if each observed variable can be reduced to a linear combination of k unobservable factors (i.e. F_k, latent variables,

Cronbach's alpha	Internal consistency
$0.9 \le \alpha$	Excellent
$0.8 \le \alpha < 0.9$	Good
$0.7 \le \alpha < 0.8$	Acceptable
$0.6 \le \alpha < 0.7$	Questionable
$0.5 \le \alpha < 0.6$	Poor
$\alpha < 0.5$	Unacceptable

Table 7.
Scales of values for the Cronbach's alpha.

considered independent from each other) and error terms (i.e. ε_i). Mathematically, that can be expressed as

$$Y_i - \mu_i = li_1F_1 + \ldots + l_{ik}F_k + \varepsilon i \quad (2)$$

In matrix terms

$$Y - \mu = LF + \varepsilon \quad (3)$$

$$\begin{bmatrix} Y_1 \\ \vdots \\ Y_i \\ \vdots \\ Y_p \end{bmatrix} - \begin{bmatrix} \mu_1 \\ \vdots \\ \mu_i \\ \vdots \\ \mu_p \end{bmatrix} = \begin{bmatrix} l_{11} & \cdots & l_{1j} & \cdots & l_{1k} \\ \vdots & \ddots & \vdots & \vdots & \vdots \\ l_{i1} & \cdots & l_{ij} & \cdots & l_{ik} \\ \vdots & \cdots & \vdots & \ddots & \vdots \\ l_{p1} & \cdots & l_{pj} & \cdots & l_{pk} \end{bmatrix} \begin{bmatrix} F_1 \\ \vdots \\ F_j \\ \vdots \\ F_k \end{bmatrix} + \begin{bmatrix} \varepsilon_1 \\ \vdots \\ \varepsilon_i \\ \vdots \\ \varepsilon_p \end{bmatrix} \quad (4)$$

Where

$Y = [Y_1, \ldots, Y_i, \ldots, Y_p]^T$ is the vector of p observable random variables.

$\mu = [\mu_1, \ldots, \mu_i, \ldots, \mu_p]^T$ is the vector of the mean values of Y.

$F = [F_1, \ldots, F_j, \ldots, F_k]^T$ is a vector of k unobserved random variables, called "common factors" as they influence all the observed Y_i.

$L = \begin{bmatrix} l_{11} & \cdots & l_{1j} & \cdots & l_{1k} \\ \vdots & \ddots & \vdots & \vdots & \vdots \\ l_{i1} & \cdots & l_{ij} & \cdots & l_{ik} \\ \vdots & \cdots & \vdots & \ddots & \vdots \\ l_{p1} & \cdots & l_{pj} & \cdots & l_{pk} \end{bmatrix}$ is a matrix of unknown constants, called "loadings" that have to be calculated.

$\varepsilon = [\varepsilon_1, \ldots, \varepsilon i, \ldots, \varepsilon_p]^T$ is a vector of unobserved stochastic error terms, with zero mean and fin te variance, that can assume different values for each i.

Assuming that:

- εi are independent from one another, and $E(\varepsilon_i) = 0$ and $Var(\varepsilon_i) = \sigma_i^2$.

- F are independent from one another, as there is no relationship between factors, and are also independent from the error terms. They are also standardized to $E(F_j) = 0$ and $Var(F_j) = 1$.
- $Cov(F) = I$ so the factors are uncorrelated.
- $k \leq p$: the number of observed variables Y_i is larger or equal to the number of common factors F_j.

Any solution for the unknown values l_{ij} of Eq. (2) or (3) with the constraints for F is defi ed as factors, and L is the loading matrix.

With these assumptions, the variance of Y_i in (2) can be calculated as

$$\begin{aligned} Var(Y_i) &= l_{i1}{}^2 Var(F_1) + ... + l_{ik}{}^2 Var(F_k) + 1^2 Var(\varepsilon_i) \\ Var(Y_i) &= l_{i1}{}^2 + ... + l_{ik}{}^2 + \sigma_i^2 \end{aligned} \tag{5}$$

where $l_{i1}{}^2 + \ldots + l_{ik}{}^2$ is the *communality* of the variance: the part that is explained by the common factors F_1, ..., F_j, ..., F_k and shared with other variables.

σ_i^2 is the *specific* variance: the part of the variance of *Yi* that is not considered in the common factors. Th s value would be equal to 0 if the common factors were perfect predictors of the observed variables.

Given two variables, Y_m and Y_n,

$$\begin{aligned} Y_m &= \mu_m + l_{m1}F_1 + \ldots + l_{mk}F_k + \varepsilon_m \\ Y_n &= \mu_n + l_{n1}F_1 + \ldots + l_{nk}F_k + \varepsilon_n \end{aligned} \tag{6}$$

The covariances can be calculated as

$$\begin{aligned} Cov(Y_m, Y_n) &= l_{m1}l_{n1}Var(F_1) + \ldots + l_{mk}l_{nk}Var(F_k) + (1)(0)\varepsilon_m + (0)(1)\varepsilon_n \\ Cov(Y_m, Y_n) &= l_{m1}l_{n1} + \ldots + l_{mk}l_{nk} \end{aligned} \tag{7}$$

Th s shows that the covariance of two variables is equal to the scalar product of their loadings.

With the expressions in Eqs. (5) and (7), it is possible to construct a *theoretical variance–covariance matrix*, implied by the model's assumptions. Then, with the data collected in the survey, an *observed variance and covariance matrix* can be calculated and constructed. If the model's assumptions are correct, it is possible to estimate the loadings l_{ij} in order to obtain a theoretical matrix closer to the observed one.

To extract the fi st set of loadings and factors from the observed variables, there are different methods. However, principal component analysis (PCA) and common factor analysis (CFA) are the most preferred and most used:

- *Principal component method*, or component factor analysis, determines the loadings l_{ij} that allows a close estimation of the total communality to the sum of the observed variances, while ignoring the covariances. The principal components are chosen by extracting the maximum variance and putting it in the fi st factor, gathering as much of the variation in the data as possible. Then, the variance explained by the fi st factor is removed and then extracts the maximum variance for the second factor; repeating the process until the last factor.

Th s model can be written as

$$C_1 = l_{11}Y_1 + l_{21}Y_2 + l_{31}Y_3 \tag{8}$$

- *Common factor analysis*: the factors are linear combinations that maximize the common portion of the variance and put them into factors, underlying *latent constructs*. Th s method does not include the specific part of the variance to determine the factor, and it is used for structural equation modeling.

Th s model can be set up as

$$\begin{aligned} Y_1 &= l_{11}F_1 + \varepsilon_1 \\ Y_2 &= l_{21}F_1 + \varepsilon_2 \\ Y_3 &= l_{31}F_1 + \varepsilon_3 \end{aligned} \tag{9}$$

Once the factors are extracted, their eigenvalues (or characteristic roots) pro- vides the amount of variance explained by every factor out of the total variance. Then, the number of factors is reduced by retaining only those which have an eigenvalue larger than 1, according to Kaiser's criterion [29].

The factor loadings obtained represent the amount of variance explained by the variable on every factor. In structural equation modeling, a value of 0.7 or higher represents that the factor extracts suffici t variance from that particular variable.

The loading values may be hard to interpret at a fi st glance. So, in a second step of the analysis, the loadings obtained can be "rotated" in order to arrive at another set of loadings, which renders the values more understandable, while fitting the observed variances and covariances equally. The effect of rotating the factors produces that each variable loads more strongly only on one of the factors and weakly on the other factors, producing the eigenvalues to vary.

There are several rotation methods that provide different solutions, arising to different interpretation. The interpretation of each factor and the number of factors needed are very subjective, and the researcher has the task to identify what is the meaning of each factor (i.e. which is the unknown latent variable hidden in the indicators).

From a general point of view, the rotation methods can be subdivided in orthogonal (when the factors cannot correlate) or oblique (the factors are allowed to correlate). The most common methods within each one of two groups are listed below:

- Orthogonal methods
    - *Varimax*: it aims to minimize the complexity of each factor by relating them to few variables while discouraging the detection of factors that influence all the variables. It produces the increase of the strongest loading values while decreasing the weaker ones in each factor.
    - *Quartimax*: it aims to fi d a general factor (or a reduced amount of them), on which most variables are loaded to, while minimizing the number of factors needed to explain each variable. Th s is done by increasing the strongest loading values while decreasing the weaker ones in each variable. Th s factor structure is usually not helpful to the research purpose.
    - *Equimax*: it is a method that attempts to simplify both factors and variables.

- Oblique methods
    - *Direct oblimin* is the standard method when the factors are allowed to be correlated, resulting in higher eigenvalues, but the interpretability of the factors may be reduced.
    - *Promax* is an alternative to the previous one, used for large dataset as it is computationally more effici t.

Example 1: the electric vehicle case study

An overview of the preliminary statistical analysis for the electric vehicle case study[1] introduced in Section 3.2 is provided. The proposed example includes analysis of Cronbach's alpha, mean and standard deviations, factor analysis with principal components as the extraction method and rotated component matrix using the Varimax method.

First of all, it may be observed that Cronbach's alpha is not consistent for all answers; therefore, the survey reliability is confi med only in the case of psychological statements referred to the environment, the technical features and the EV's advantages (i.e. Cronbach's alpha is higher than 0.5). Among them, a further analysis is provided in terms of mean and standard deviations: with respect to the attitude about the environment, for both statements the mean is higher than 3, and the standard deviation is lower than 1; regarding the attitude about technical features, higher values of mean (higher than 3) and lower values (lower than 1) of standard deviations are observed for the following statements (refer to Section 3.2 for the meaning of the following variables): <*F_tech_fea*>, <TF*_power* > and < TF*_range*>. Finally, the perceptions of EV's advantages show mean values higher than 3 and standard deviations lower than 1 only for the following statements: <AD_{Red_CO2}>, <$AD_{Efficiency}$ > and < AD_{Red_poll}>. However, even though the Cronbach's alpha is not satisfying for the perceptions of disadvantages of EVs, the mean values and the standard deviations for all statements (<DIS_{infr}>, <DIS_{red_fea}>, <DIS_{batt_range}>) highlight their relevance on users' behaviour. All signifi ant results are in bold in **Table 8**.

Attitudes about environment		
Cronbach's alpha = 0.551		
	mean	*SD*
F_cons	4.32	0.71
F_poll	3.13	0.98
Attitudes about technical features		
Cronbach's alpha = 0.656		
	mean	*SD*
F_tech_fea	3.68	0.85
TF_power	3.51	0.84
TF_top_speed	2.98	0.85
TF_accel	3.14	1.00
TF_range	4.53	0.73
Perceptions of advantages of EVs		
Cronbach's alpha = 0.549		
	Mean	*SD*
AD_{Red_CO2}	4.11	0.82
$AD_{Efficiency}$	3.42	1.08

AD_{Red_poll}	3.62	0.97
AD_{Less_parts}	3.01	1.14
Perceptions of disadvantages of EVs		
Cronbach's alpha = 0.364		
	Mean	SD
DIS_{infr}	4.21	0.82
DIS_{red_fea}	3.25	0.94
DIS_{batt_range}	4.05	0.87

Table 8.

Mean and standard deviations of the responses in the EV study.

Factor analysis				
Variables		***Components***	***Rotated components***	
	1	2	1	2
AD_red_CO_2	0.721	-0.306	0.773	0.125
AD_red_poll	0.601	-0.018	0.751	-0.156
AD_effici cy	0.553	-0.532	0.519	0.305
F_consumption	0.548	0.518	-0.068	0.768
F_pollution	0.352	0.686	0.188	0.731

Table 9.

Factor analysis of the EV study.

The factor analysis carried out through the principal component analysis extraction method, allowed to identify the latent factors correlated to the psychological statements. The components extracted were also rotated using the Varimax method.

The results (signifi ant values are highlighted in bold in **Table 9**) underline the correlations among the following statements referred to the advantage perceptions <AD_{Red_CO}>, <$AD_{Efficiency}$ > and < AD_{Red_poll} > and all statements regarding the environmental attitude, <*F_consumption* > and < *F_pollution* > .

Mathematical formulation

Choice function and structural and measurement equations

The utility choice function in the hybrid choice model is based on the assumption that each individual is faced with a set of alternatives, *i*, and each alternative expressed as a function of a vector of observed instrumental attributes, *Xi*; the users' attributes, Xi,SE; a vector of latent variables, *LVi*; and the error term *εi:*

$$U^i = \beta_x X^i + \beta_{SE} X^i_{SE} + \beta_{LV} LV^i + \varepsilon^i \quad (10)$$

---

[1] For the sake of brevity only, the results which are related to one case study, the electric vehicle case study, are displayed.

With reference to the LV^i vector, two equations have to be specifi d: the structural and the measurement equations. The structural equations are introduced in order to specify the latent variables, while the measurement equations are introduced in order to specify the perception indicators.

In particular, if p is the generic latent variable, the structural equation for each latent variable may be expressed as follows:

$$LV_p^i = \gamma_p + \sum_j \beta_{SE,j} X_{SE,j}^i + \omega_p^i \tag{11}$$

where γ_p is the intersect, $X^i_{SE,j}$ is the vector of the users' characteristics attributes, $\beta_{SE,j}$ is the vector of the coeffici ts associated with the users' characteristics (to be estimated), ω^i_p is the error term which is usually normally distributed with zero mean and $\sigma_{\omega,p}$ is the standard deviation.

Furthermore, let I^i_n be a vector of perceptions indicators associated to each latent variable. Each perception indicator (i.e. vector component) may be specifi d by a measurement equation as follows:

$$I_{p,k}^i = \alpha_{p,k} + \lambda_{p,k} LV_p^i + \nu_{p,k}^i \tag{12}$$

where $\alpha_{p,k}$ is the intersect, $\lambda_{p,k}$ is the coeffici t associated with the latent variable (to be estimated), ν^i_{pk} is the error terms usually assumed normally distributed with zero mean and $\sigma_{\nu pk}$ is the standard deviation of the error term.

The psychometric indicators that reveal the latent variables may be coded using a Likert scale [19]. These indicators can be considered to be a linear continuous expression of the LV's or an ordered discrete variable. The fi st approach has been historically chosen because simpler and more practical with lower computational cost. However, assuming these indicators as continuous variables are in contrast with the real nature of the Likert scale (the Likert scale is a discrete measure) [30], such an approach may introduce some biases in the parameters' estimation. In recent years, several studies have treated them as discrete variables, but with a higher computational cost [31]. In particular, if the measurement is represented by an ordered discrete variable J taking the values $j_1, j_2, \ldots, j_M$, we have

$$\mathrm{J} = \begin{cases} j_1 & if\ \mathrm{I} < \tau_1 \\ j_2 & if\ \tau_1 \leq \mathrm{I} < \tau_2 \\ \vdots & \\ j_i & if\ \tau_{i-1} \leq \mathrm{I} < \tau_i \\ \vdots & \\ j_M & if\ \tau_{M-1} \leq \mathrm{I} \end{cases} \tag{13}$$

where I is defi ed by Eq. (12) and $\tau_1, \ldots, \tau_{M-1}$ are parameters to be estimated, such that

$$\tau_1 \leq \tau_2 \leq \ldots \leq \tau_i \leq \ldots \leq \tau_{M-1}$$

If the measurements use a Likert scale with M = 5 levels, four parameters τi are needed. But, in order to account for the symmetry of the indicators, two positive parameters δ_1 and δ_2 are specifi d instead, in order to defi e

$$\begin{aligned} \tau_1 &= -\delta_1 - \delta_2 \\ \tau_2 &= -\delta_1 \\ \tau_3 &= \delta_1 \\ \tau_4 &= \delta_1 + \delta_2 \end{aligned} \tag{14}$$

Then, the probability of a given response j_i is given by the ordered probit model [5]. For completeness, in the following section, the estimation results related to the

HySolarKit and the electric vehicle case study are shown.

In this research report, the model parameters were estimated in accordance with the maximum simulated likelihood statistical approach.

Example: the HySolarKit case study

The fi st results shown in this section refer to the HySolarKit case study. As already anticipated in Section 3.1, the choice set was composed of two alternatives: "install" and "not-install". In the following the estimation results are presented, distinguishing the choice utility function, the structural equations and the measurement equations.

Parameters of the choice utility function

The utility choice functions were analytically specifi d in accordance with the following equation:

$$U^i = \beta_x X^i + \beta_{SE} X^i_{SE} + \beta_{LV} LV^i + \varepsilon^i \tag{15}$$

The results are shown in **Table 10.** In particular, the estimation results underline the following latent variables as statistically signifi ant: attitudes towards fuel consumption (LV1), towards the vehicle design (LV2) and towards the environment (LV3).

The coeffici ts related to the parameters in the measurement equation for an ordinal specifi ation are estimated in the considered model. As the measurements are using a Likert scale with seven levels, six parameters τ_i are needed

Attributes	*Attributes coefficients (betas)*
	Install Not-install
Age *[def: age of the respondent]*	+0.160 *(+0.960)*
Master's degree *[def: 1 for users achieved this educational attainment]*	+ 0.156 *(+1.16)*
ZonRes *[def: 0 for users living to the historical centre, 1 if in the outskirts]*	+0.0761 *(+0.960)*
CarAge *[def: age of the owned car on which the respondent would install the kit]*	+0.0272 *(+1.55)*
by Car-Shopping *[def: mode choice "Car" and trip purpose "Shopping"]*	+0.669 *(+1.490)*
by Car-Personal Services *[def: mode choice "Car" and trip purpose "Personal Services"]*	+0.192 *(+0.53)*

Δcost *[def: In order to compare the scenarios with and without the kit, the financial gain was expressed in terms of weekly costs, rather than yearly costs or life costs based on the year of the vehicle owned by the respondent]*	+0.0638 *(+8.16)*
LV_1 *[def: latent variable representing consumption attitude]*	+0.548 *(+2.55)*
LV_2 *[def: latent variable representing design attitude]*	+0.0682 *(+0.46)*
LV3 *[def: latent variable representing environment attitude]*	+0.104 *(+0.98)*
Statistics	
Number of respondents	1364
Number of observations	1364
Init-log-likelihood1	-944.760
Final log-likelihood	-779.81
Rho-square	0.212

**in parenthesis the t-test values.*
1Only the log-likelihood associated with the discrete choice component is considered.

Table 10.
Attribute coefficients of the choice model. HySolarKit case study.

$DELTA_1$	+ 1.46 *(+ 38.90)*
$DELTA_2$	+ 1.34 *(+ 43.85)*
$DELTA_3$	+ 1.52 *(+ 46.11)*

**in parenthesis the t-test values.*

Table 11.
Delta values of the calibrated measurement equations. HySolarKit case study.

(see Section 5.1). However, in order to account for the symmetry of the indicators, three positive parameters δ_1, δ_2 and δ_3 are actually required (**Table 11**).

Parameters of the structural model

The coeffici ts in the structural model are analytically represented by the following equation:

$$LV_p^i = \gamma_p + \sum_j \beta_{SE,j} X_{SE,j}^i + \omega_p^i \quad (16)$$

Th s equation shows that each latent variable is a function of an intercept value γ_p of beta-coeffici ts $\beta_{SE,j}$ for each of the socioeconomic attributes $X^i_{SE,j}$ of the respondents that influence the latent variable and contains an error term ω^i_p normally distributed with zero mean and $\sigma_{\omega p}$ standard deviations.

The estimation results displayed in below refers to the signifi ant latent variables of the model, representing the attitude towards the fuel consumption (LV_1), the vehicle design (LV_2) and the environment (LV_3) (**Table 12**).

Parameters of the measurement model

Finally, with regard to the measurement model depending on the perception indicators, they are analytically represented by the following equation:

$$I^i_{p,k} = \alpha_{p,k} + \lambda_{p,k} LV^i_p + \nu^i_{p,k} \tag{17}$$

where each perception indicator is a function of an intercept value $\alpha_{p,k,}$ a coeffici t $\lambda_{p,k}$ associated with the latent variable and an error term $\nu^i_{p,k}$ assumed normally distributed with zero mean and $\sigma_{\nu pk}$ standard deviation.

Table 13 shows the coeffici ts for each perception indicator, which were specifi d in accordance with the preliminary analyses (not shown for the sake of brevity). In particular, the fi st latent variable about fuel consumption (LV_1) is described by two indicators <I_{cons0} > and < I_{cons2}>. The second latent variable, about vehicle design (LV_2), is described by four perception indicators, <$I_{design0}$>, <$I_{design1}$>, <$I_{design3}$ > and < $I_{design4}$>. Finally, the last latent variable representing the attitudes towards the environment (LV_3) is described by four indicators < I_{env0}>, < I_{env3}>, < I_{env4} > and < I_{env6} > .

In general, the estimation results underline the necessity to introduce two different kinds of questions: direct and indirect questions.

Structural model	
Attributes	***Attribute coefficients (betas)***
LV1: Attitude towards fuel consumption	
γ_1	-2.32 *(-29.80)*
ω_1	+0.787 *(+16.37)*
by car-shopping *[def: mode choice "Car" and trip purpose "Shopping"]*	-0.448 *(-2.52)*
by car-personal services *[def: mode choice "Car" and trip purpose "Personal Services"]*	+0.550 *(+3.88)*
Master's degree *[def: 1 for users achieved this educational attainment]*	+0.128 *(+2.22)*
LV_2: Attitude towards design issues	
γ_2	-3.39 *(-42.17)*
ω_2	+0.299 *(+6.96)*
Age *[def: age of the respondent]*	-0.0955 *(-2.37)*
SE_male *[def: respondent's gender (0 = female; 1 = male)]*	-0.278 *(-6.65)*
LV_3: Attitude towards environment	
γ_3	—

ω_3	+1.34 *(+27.18)*
Age *[def: age of the respondent]*	*-1.06* *(-15.60)*
SE_male *[def: respondent's gender (0 = female; 1 = male)]*	*-1.12* *(16.74)*
**in parenthesis the t-test values.*	

Table 12.
Coefficients of the calibrated structural model. HySolarKit case study.

Example 2: the electric vehicle case study

As described in Section 3.2, the considered choice set is composed of two alternatives: the alternative A representing the respondent's willingness to buy an electric vehicle and the alternative B corresponding to the willingness of buying a conventional (diesel) vehicle. The provided results are referred to an ordered model specifi ation.

In the following the estimation results are presented, distinguishing the choice of utility function, the structural equations and the measurement equations.

Parameters of the choice utility function

The attribute coeffici ts in the utility choice function are analytically specifi d by the following equation:

$$U^i = \beta_x X^i + \beta_{SE} X^i_{SE} + \beta_{LV} LV^i + \varepsilon^i \tag{18}$$

Measurement model					
Fuel consumption		***Vehicle design***		***Environment***	
I_{cons0} *[def: The vehicle fuel consumption significantly influences my choice in purchasing a new car]*		$I_{design0}$ *[def: The vehicle design significantly influences my choice in purchasing a new car]*		I_{env0} *[def: The evaluation of the environmental impact significantly influences my choice in purchasing a new car]*	
α_{10}	+0.567 (+3.46)	α_{20}	−1.79 (−2.77)	α_{30}	−1.89 (−24.48)
λ_{10}	+0.725 (+11.40)	λ_{20}	+0.148 (+0.85)	λ_{30}	+0.495 (+14.12)
ν_{10}	+0.787 (+16.37)	ν_{20}	+1.38 (+33.16)	ν_{30}	+1.18 (+31.06)
I_{cons2} *[def: I am usually attentive to the special offers of electric operators]*		$I_{design1}$ *[def: When parking I am usually careful to avoid having my car damaged]*		I_{env3} *[def: I really enjoy spending my free time in parks and green areas to breathe clean area]*	
α_{12}	0	α_{21}	0	α_{33}	−1.36 (−19.03)

λ_{12}	1	λ_{21}	1	λ_{33}	+0.729 *(+20.82)*
ν_{12}	1	ν_{21}	1	ν_{33}	+0.983 *(+25.99)*
		$I_{design3}$ *[def: When furnishing I am willing to buy pieces with modern design features and original details]*		I_{env4} *[def: How much do you agree with following sentence: We must act and make decisions to reduce emissions of greenhouse gases]*	
		α_{23}	+2.79 *(+2.72)*	α_{34}	0
		λ_{23}	+1.60 *(+5.72)*	λ_{34}	1
		ν_{23}	+1.43 *(+31.24)*	ν_{34}	1
		$I_{design4}$ *[def: I am willing to go to the body shop mechanic not only for major damages]*		I_{env6} *[def: I am not willing to use the car during weekend to protect the environment and then reduce air pollution]*	
		α_{24}	+2.79 *(+2.69)*	α_{36}	−1.95 *(−26.15)*
		λ_{24}	+1.84 *(+6.52)*	λ_{36}	+0.396 *(+12.18)*
		ν_{24}	+1.65 *(+28.17)*	ν_{36}	+1.13 *(+31.68)*

[*]in parenthesis the t-test values.

Table 13.

Coefficients of the calibrated measurement model. HySolarKit case study.

The results are shown in **Table 14.** In particular, the following attitudes were statistically signifi ant: the attitudes towards the environment (LV_1) and the perception of the advantages of EVs (LV_2).

Attributes	***Attribute coefficients (betas)***	
	BUY - EV	***BUY - CV***
VAR_monthly_cost_abs *[def: Variation in monthly cost [EUR] between an electric car and a conventional one]*		+0.118 *(+17.65)*
SE_AutoSI *[def: 1 if the respondent has at least 1 car in the household]*		+0.865 *(+2.02)*
F_tech_fea [5] *[def: The vehicle technical features significantly influence my choice in purchasing a new car (5 = strongly agree)]*		+0.451 *(+1.67)*
DIS_red_fea [5] *[def: Compared to a normal car, EV are inferior in terms of performances (5 = strongly agree)]*		+0.778 *(+2.25)*

LV_1	+2.10
	(+11.86)
LV_2	+0.435
	(+1.57)
Statistics	
Number of respondents	1462
Number of observations	1462
Init-log-likelihood[1]	−1013.38
Final log-likelihood	−385.55
Rho-square	0.620

*in parenthesis the t-test values.
[1]Only the log-likelihood associated with the discrete choice component is considered

Table 14.
Attribute coefficients of the choice model. EV case study.

$DELTA_1$	0.531
	(+30.25)
$DELTA_2$	1.27
	(+38.94)

*in parenthesis the t-test values.

Table 15.
Delta values of the calibrated measurement equations. EV case study.

The coeffici ts related to the parameters in the measurement equation for an ordinal specifi ation are estimated in the considered model. As the measurements are using a Likert scale with five levels, four parameters τi are needed. However, in order to account for the symmetry of the indicators, two positive parameters δ_1 and δ_2 are actually required (**Table 15**).

Parameters of the structural model

The coeffici ts in the structural model are analytically represented by the following equation:

Structural model	
Attributes	***Attribute coefficients (betas)***
LV_1: Attitude towards the environment	
γ_1	+1.83
	(+33.70)
ω_1	+0.119
	(+3.84)
SE_male *[def: Interviewee's gender (0 = female; 1 = male)]*	−0.242
	(−5.41)
LV_2: Perception of EV's advantages	

γ_2	+0.857
	(+11.57)
ω_2	+0.584
	(+9.86)
SE_male *[def: Interviewee's gender (0 = female; 1 = male)]*	−0.0445
	(−1.01)
SE_delta_age *[def: Age class of individuals (0 = 19 years; 1 = 20 years; 2 = 21 years; 3 = 22 years; 4 = 23 years; 5 = 24 years; 6 = 25 years; 7 = 26 years)]*	−0.0246
	(−1.39)
**in parenthesis the t-test values.*	

Table 16.
Coefficients of the calibrated structural model. EV case study.

$$LV_p^i = \gamma_p + \Sigma_j \beta_{SE,j} X_{SE,j}^i + \omega_p^i \tag{19}$$

The estimation results displayed in **Table 16** refer only to the two signifi ant latent variables of the model, standing for the attitudes towards the environment (LV_1) and the perception of the advantages of EVs (LV_2). In particular, for each latent variable, the table displays the results of the intercept value γ_p, the betacoefficients $\beta_{SE,j}$ of the socioeconomic attributes $X^i_{SE,j}$ of the respondents that influence the latent variable and the error terms ω^i_p normally distributed with zero mean and $\sigma_{\omega p}$ standard deviations of the error term.

Parameters of the measurement model

Finally, the measurement model depending on the perception indicators is analytically represented by the following equation:

$$I_{p,k}^i = \alpha_{p,k} + \lambda_{p,k} LV_p^i + \nu_{p,k}^i \tag{20}$$

These parameters were specifi d in accordance with the preliminary analyses. In particular, the fi st latent variable about the environment (LV_1) is described by two indicators, <*F_cons* > and < *F_poll*>, while the second latent variable, perception of EV's advantages (LV_2), is described by three perception indicators <AD_{Red_co2}>, <$AD_{Efficiency}$ > and < AD_{Red_poll} >. In **Table 17**, the intercept value $\alpha_{p,k}$, the

Measurement model			
LV 1: Environment		*LV 2: Perception of EV's advantages*	
F_cons *[def: The vehicle fuel consumption significantly influences my choice in purchasing a new car]*		*AD_{Red_CO2}* *[def: I am interested in EV to contribute to the emissions reduction]*	
α_{10}	−4.62 (−4.92)	α_{20}	+0.391 (+2.96)
λ_{10}	+2.85 (+5.23)	λ_{20}	+1.32 (+8.40)
ν_{10}	+1.03 (+30.15)	ν_{20}	+0.742 (+11.55)

F_poll *[def: I care about the amount of pollution generated by a car when it's being used]*		$AD_{Efficiency}$ *[def: Compared to a normal car, EV are superior in terms of energy efficiency]*	
α_{11}	0	α_{21}	−0.143 *(−1.37)*
λ_{11}	1	λ_{21}	+0.881 *(+7.27)*
ν_{11}	1	ν_{21}	+1.22 *(+27.41)*
		AD_{Red_poll} *[def: I believe using EV can significantly reduce the acoustic pollution in cities]*	
		α_{22}	0
		λ_{22}	1
		ν_{22}	1
**in parenthesis the t-test values.*			

Table 17.
Coefficients of the calibrated measurement model. EV case study.

coeffici t $\lambda_{p,k}$ associated with the latent variable and the error terms $v^{i}_{p,k}$ assumed normally distributed with zero mean and σ_{vpk} standard deviation are displayed for each perception indicator.

Conclusions

Depending on the context, several factors may affect users' choices. In this chapter, the main focus refers to modeling users' propensity to choose/adopt a new/ innovative technology. Th s is a crucial task in order to increase the attractiveness of strategies that may be employed to achieve sustainable transportation. In particular, two related main issues are still open in the literature: (a) interpreting and modeling users' behaviour towards these new technologies and (b) assessing the potential environmental impacts. It is widely recognized that traditional approaches used to interpret and model users' choice behaviour may lead to neglect the numerous nonquantitative factors that may affect users' behaviors. Indeed, users' choices may be influenced by social and psychological factors, symbolic and affective factors, habits and the confli t between collective and individual interests (e.g. car use as a commons dilemma). These imply that changes in transportation modes may be achieved either by influencing individual motivations and perceptions (*psychological strategies*) or by changing the context in which decisions are made (*structural strategies*).

The book chapter provides an overview of the methodology to be adopted in order to support psychological-based strategies. In fact, psychological factors, such as attitudes, concerns and perceptions, may play a signifi ant role which should be explicitly modeled.

Although several approaches may be identifi d in the literature able to address the above-mentioned issue, the hybrid choice modeling approach based on RUT can be considered a proper solution to explicitly consider the perceptions, attitudes and concerns in the modeling of the choice behaviour. The specifi ation of such models requires a careful survey design, rigorous preliminary descriptive analyses and the model parameter estimation. The present chapter deals with all the cited issues, fi st introducing the main criticalities in modeling choice behaviour in new technological contexts, then proposing a methodological framework and fi ally introducing different explanatory examples on real case studies. In particular, Section 3 focuses the attention on the different approaches to collect users' attitudes/perceptions and concludes the need for a mixed approach based on both direct and indirect questioning. Section 4 introduces the methodology to properly evaluate the consistency of the dataset and the latent variables identifi ation. It evidences the need for basic analysis, such as the estimation of the mean and standard deviations, and the importance of the Cronbach's alpha test and the principal components and the rotate component matrices for the identifi ation of the latent variables. Section 5 deals with the model's specifi ation issues, pointing out the most robust approach for the specifi ation and calibration of a hybrid choice model with latent variable. All the introduced sections are supported by real experimental results [24, 25] for explanatory and guideline purposes.

Acknowledgements

Th s research was partially supported by the University of Salerno, under local grant no. ORSA171328—2017.

Conflict of interest

No potential confli t of interest was reported by the authors.

Author details

Stefano de Luca*, Roberta Di Pace and Facundo Storani
University of Salerno, Fisciano, Salerno, Italy

*Address all correspondence to: sdeluca@unisa.it

References

[1] Bahamonde-Birke FJ, Kunert U, Link H, de Ortúzar JD. About attitudes and perceptions: Finding the proper way to consider latent variables in discrete choice models. Transportation. 2017; 44(3):475-493. Available from: http:// link.springer.com/10.1007/s11116-015- 9663-5

[2] Manski CF. The structure of random utility models. Theory and Decision. 1977;8(3):229-254

[3] Manski CF, McFadden D. Alternative estimators and sample designs for discrete choice analysis. Structural Analysis of Discrete Data with Econometric Applications. 1981;2:2-50

[4] Ben-Akiva ME, Lerman SR. Discrete Choice Analysis: Theory and Application to Travel Demand. Vol. 9. London, England: MIT Press; 1985

[5] Train KE. Discrete Choice Methods with Simulation. Cambridge: Cambridge University Press; 2003. Available from: http://ebooks.cambridge.org/ref/id/ CBO9780511753930

[6] Daziano RA, Bolduc D. Incorporating pro-environmental preferences towards green automobile technologies through a Bayesian hybrid choice model. Transportmetrica A: Transport Science. 2013;9(1):74-106

[7] Ben-Akiva M, McFadden D, Gärling T, Gopinath D, Walker J, Bolduc D, et al. Extended framework for modeling choice behavior. Marketing Letters. 1999;10(3):187-203. Available from: https://link.springer. com/article/10.1023%-2FA% 3A1008046730291

[8] Walker J. Extended discrete choice models: Integrated framework, flex ble error structures, and latent variables. Environmental Engineering. PhD Thesis. Massachusetts Institute of Technology. 2001

[9] McFadden D. Economic choices. The American Economic Review. 2001; 91(3):351-378

[10] Ben-Akiva M, Mcfadden D, Train K, Walker J, Bhat C, Bierlaire M, et al. Hybrid choice models: Progress and challenges. Marketing Letters. 2002; 13(3):163-175

[11] Walker J, Ben-Akiva M. Generalized random utility model. Mathematical Social Sciences. 2002;43(3):303-343. Available from: http://linkinghub. elsevier.com/retrieve/pii/ S0165489602000239

[12] Ortúzar J de D, Hutt GA. La influencia de elementos subjetivos en funciones desagregadas de elección discreta. Ingeniería de Sistemas. 1984; 4(2):37-54

[13] McFadden D. The choice theory approach to market research. Marketing Science. 1986;5(4):275-297. Available from: http://pubsonline.informs.org/d oi/abs/10.1287/mksc.5.4.275

[14] Jöreskog KG. A general method for estimating a linear structural equation system. ETS Research Bulletin Series. 1970;1970(2):i-41. Available from: h ttp://doi.wiley.com/10.1002/ j.2333-8504.1970.tb00783.x

[15] Raveau S, Álvarez-Daziano R, Yáñez MF, Bolduc D, de Dios Ortúzar J. Sequential and simultaneous estimation of hybrid discrete choice models. Transportation Research Record. 2010; 2156(1):131-139. Available from: http:// journals.sagepub.com/doi/10.3141/2156-15

[16] Bolduc D, Alvarez-Daziano R. On estimation of hybrid choice models. In: Choice Modelling: The State-of-the-Art and the State-of-Practice. United Kingdom: Emerald Group Publishing Limited; 2010. pp. 259-287. Available from: http://www.emeraldinsight.com/ doi/10.1108/9781849507738-011

[17] de Almeida Correia GH, de Abreu e Silva J, Viegas JM. Using latent attitudinal variables estimated through a structural equations model for understanding carpooling propensity. Transportation Planning and Technology. 2013;36(6):499-519. Available from: http://www.tandfonline.com/doi/abs/10.1080/ 03081060.2013.830894

[18] Louviere JJ, Hensher DA, Swait JD. Stated Choice Methods: Analysis and Applications. United Kingdom: Cambridge University Press; 2000

[19] Likert R. A technique for the measurement of attitudes. Archives of Psychology. 1932;22(140):5-55. Available from: https://legacy.voteview. com/pdf/Likert_1932.pdf

[20] Allport GW. Attitudes. In: A Handbook of Social Psychology. 1935. pp. 798–844

[21] Pickens J. Attitudes and perceptions. Organizational Behavior in Health Care. 2005:43-76

[22] Norman DA, Lindsay PH. Human Information Processing: An Introduction to Psychology. Elsevier: Academic Press; 1977. Available from: h ttps://linkinghub.elsevier.com/retrieve/ pii/C2013011094X

[23] de Luca S, Di Pace R, Marano V. Modelling the adoption intention and installation choice of an automotive after-market mild-solar-hybridization kit. Transportation Research Part C: Emerging Technologies. 2015;56: 426-445. Available from: https:// linkinghub.elsevier.com/retrieve/pii/S0968090X15001643

[24] de Luca S, Di Pace R. Attitudes did predict "better" choice behaviour towards AFV (new tech): The hysolarkit case study. Under Rev Sustain

[25] de Luca S, Di Pace R, Storani F. The impact of attitudes and perceptions in modelling users' willingness to purchase electric vehicles in immature markets. Transportation Research Part A

[26] Rizzo G, Marano V, Pisanti C, D'Agostino M, Naddeo M, Sorrentino M, et al. A prototype mild-solar- hybridization kit: Design and challenges. Energy Procedia. 2014;45:1017-1026. Available from: https://linkinghub. elsevier.com/retrieve/pii/ S1876610214001088

[27] de Luca S, Di Pace R. Aftermarket vehicle hybridization: Potential market penetration and environmental benefits of a hybrid-solar kit. International Journal of Sustainable Transportation. 2018;12(5):353-366. Available from: h ttps://www.tandfonline.com/doi/full/ 10.1080/15568318.2017.1377325

[28] de Luca S, Di Pace R, Storani F. A study on users' behaviour towards electric vehicles in immature markets: The Argentina case study. In: 2018 IEEE International Conference on Environment and Electrical Engineering and 2018 IEEE Industrial and Commercial Power Systems Europe (EEEIC/I&CPS Europe). IEEE; 2018. pp. 1–6. Available from: https://ieee xplore.ieee.org/document/8493840/

[29] Kaiser HF. The application of electronic computers to factor analysis. Educational and Psychological Measurement. 1960;20(1):141-151. DOI: 10.1177/001316446002000116

[30] Daly A, Hess S, Patruni B, Potoglou D, Rohr C. Using ordered attitudinal indicators in a latent variable choice model: A study of the impact of security on rail travel behaviour. Transportation. 2012;39(2):267-297. Available from: https://link.springer.com/article/ 10.1007/s11116-011-9351-z

[31] Bahamonde-Birke FJ, Ortúzar J de Analyzing the continuity of attitudinal and perceptual indicators in hybrid choice models. Journal of Choice Modeling. 2017;25:28-39. Available from: https://linkinghub.elsevier.com/ retrieve/pii/S1755534517300118

9

Railway Infrastructure Capacity in the Open Access Condition: Case Studies on SŽDC and ŽSR Networks

Jozef Gašparík and Václav Cempírek

Abstract

The railway sector in the European Union is changing. The goal of EU transport policy is to liberalize the market for rail transport services, dismantle national transport monopolies, and open competitive public tenders to other train operators. For the optimal utilization of the railway infrastructure capacity, it is necessary to calculate it properly in terms of open access to the infrastructure. At present, many important corridors are at full capacity. Therefore, in order to increase the number of freight trains, it is necessary to implement certain measures to increase the track line capacity. Infrastructure capacity research is part of the complexity of the capacity management processes. A progressive approach to defi e it means to describe the estimating process of railway infrastructure capacity including progressive capacity allocation approaches as a key part of capacity management. The aim is to defi e the processes of the infrastructure capacity management on which depends the quality level of operational traffi management as well the effici cy of the traffi fl w on the infrastructure. The partial objective is to investigate the impact of systematic train paths in periodic timetables on rail infrastructure capacity. The proposals fully respect the EU transport policy.

Keywords: transport policy, rail transport market, infrastructure capacity, train path, open access, capacity measurement

Introduction

The restructuring of the rail market has created new relationships between the players in this market. The examination of the relationship between the entities, the infrastructure manager, and the railway undertaking (herewith referred to as the RU) focuses on the assessment of the allocation of railway infrastructure capacity, with a view to knowing traffi technology and track-side technologies including the economic aspects. Th s complex issue is closely related to the determination of the capacity of the railway infrastructure, which represents the maximum possible offer of train paths for the infrastructure manager to construct the timetable [1–3].

The timetable is the operating plan and also offers train paths to potential customers. Loss of stability and timetable quality may result in the absence of spare capacity, that is, its exhaustion after allocation and after the completion of the timetable for the scheduled period [4].

The main objective of this chapter is to defi e the processes of managing the capacity of railway infrastructure with the aim of achieving high-quality operative management of traffi due to the effici cy of transport fl w on the infrastructure. These objectives fully respect EU transport policy, which provides a framework for creating transparent conditions and minimizing risks in accessing transport infrastructure and ensuring the growing transport needs of the company at the required time and quality. The scientific contribution is proven in the application of theoretical knowledge in the fi ld of railway transport technology in terms of a case study on the corridor's lines.

Requirements of carriers for railway infrastructure capacity

The allocation of railway infrastructure capacity is a complex product of the infrastructure manager, which consists of a number of sub-services. The infrastructure manager is obliged to publish the conditions of access to the infrastructure (the so-called Network Statement [5]) and to determine the free capacity of each line section. Consequently, it is intended to allow nondiscriminatory access for railway undertakings.

The Network Statement contains mainly the technical characteristics of the railways, conditions for the allocation of rail capacity to applicants, including procedures for the lack of rail capacity, conditions for access to the network, information on the price for the allocation of rail capacity and pricing for the use of the infrastructure, requirements for the application for the allocation of rail capacity, etc.

The allocation of capacity is the sale of a particular train path(s) on specific line sections in a specific time window. From a technological point of view, it is important to correctly determine the technical capacity of the track section, that is, to determine the extent of train traffi for a given track section, to show suffici t stability of train traffi even during operational irregularities. The charging of capacity is one of the tools for number of train path regulation [6–8]. The economic aspect of capacity takes into account also the risk of paying sanctions to railway undertakings for failure to comply with the RU due to poor organization of train transport (see the introduction of the European Performance Regime (EPR) [9]). The process demands the reconciliation of the requirements of all RU with regard to the technological nature of rail transport, requiring railway undertakings to anticipate traffic ws and commodity fl ws in the medium term.

The line capacity, that is, the ability to insert the required train paths for a given part of the infrastructure in a certain time period, is expressed by the number of train paths that can be determined over a certain time window with given technical, operational, and personnel equipment and with the necessary transport quality achieved [1, 10, 11].

Capacity defin tion in UIC 406 Capacity [12] represents some consensus among individual infrastructure managers on this specific issue and suggests that a clear defin tion of capacity cannot be established. The International Railway Union (UIC) defi es railway infrastructure capacity as "the total number of possible paths in a defi ed time window, considering the actual path mix or known developments, respectively, and the IM's own assumptions; in node individual lines or part of the network, with market-oriented quality" [1].

In principle, capacity (throughput permeability) can be determined by the following approaches [12, 10]:

- graphically;
- graphically-analytically;
- analytically; or
- simulation modeling.

An overview of the most preferred methodologies for capacity estimation is elaborated in the work [13]. The most comprehensive approach is to use simulation tools to evaluate the capacity of a railway infrastructure based on a real-world traffi model. Capacity assessment using simulation modeling methods provides a comprehensive assessment of the capacity characteristics of the transport infrastructure being solved. The result provided is only suboptimal in terms of the general approach depending on the course of the simulation. A problem here is the range of input data required for a simulation model (a detailed description of the infrastructure and dynamic properties of the vehicles), as well as the time data required for the simulation assessment. On the other hand, the new possibilities offered by simulation modeling are a prerequisite for its successful implementation in cases where it is justifi d. The criterion used here is primarily the stability of the timetable (the ability not to increase or decrease the input delay).

Relevant data for capacity allocation in the case study

The capacity assessment is based on the evaluation of the existing timetable. Infrastructure dimensioning, operational performance, and quality of service are interdependent. If two variables are known, the third can be derived. Security requirements, general economic framework conditions, and environmental constraints are given by the external environment.

The following factors influence the capacity of a given infrastructure [11, 14]:

- number of train paths over a specifi d time interval;
- average speed;
- stability of train traffic bility to dispose of the initial delay and its nontransfer to other trains); and
- heterogeneity, that is, with the number of different driving times and their large differences, capacity utilization increases.

The overview of the preferred methodologies among European Infrastructure managers is elaborated in the work of Kontaxi and Riccci [13]. The resulting average value of the stability coeffici t (ratio of the output and input delays of the train on the monitored infrastructure in the simulation run) is the basis for assessing whether the infrastructure under investigation corresponds to the expected traffic nge.

The train path is defi ed for the purposes of EP and ER 2012/34/EU directive establishing a single European railway area for the allocation of railway infrastructure capacity [15] as "the infrastructure capacity needed to run a train between two places over a given period." The train path is defi ed by important parameters, such as train type, days of operation, routing, scheduled speed, arrival times, departures times, and transfer times at stations and stops.

The timetable shows the paths of all regular trains, trains as needed (on days of deployment), and canceled trains (they travel on specifi d days and their paths cancel other train paths where the regular train must be waiting in the event of the jamming train being introduced). The insertion of train paths to the graphical timetable must be in accordance with the technological procedures of the station operation processes and traffi safety. The insertion of train paths to the timetable under SŽDC and ŽSR Railway infrastructure manager is performed gradually according to the following basic types [5, 11]:

- international and national expresses and fast trains;
- long-distance passenger trains;

- freight express trains;
- passenger trains to and from employment;
- feeder freight service trains on line sections before determining the fi al position of the running freight trains; and
- interfering train paths (regular train paths with special path construction, these trains are operated in dedicated days during the week or daily during a specific ime period).

The RU submits timetable capacity requirements along with mandatory data summarized in **Table 1**. Only three of the twelve required data affect the calculation of train running times and thus the line capacity. At the same time, the RU communicates other specific data whose operational nature allows the train to assign the particular type and traffi calendar in particular. Although this is not data directly affecting infrastructure utilization, it is data that allow the capacity allocator to decide on the allocation or nonallocation of railway capacity from a legislative perspective.

Required data under the Network Statement	Influence on driving time
Identifi ation data of the RU	N
Position route guidance	N
Timing of train path	N
Train type	N
Train set data specifi ations	Y
Technical data on traction vehicles including their number and function in the train	Y
Train driving calendar	N
Type of railway transport operated	N
Required tariff nd nontariff c nditions	N
Type and extent of services provided in the train	N
RU requirements for technological operations in stations	Y
Known extraordinary of the train	N
Source: authors on the ground of [4].	

Table 1.
Required data in rail capacity allocation request.

Th ee requirements summarize the kit's technical data, that is, the transport weight and the length of the train, tractive vehicle order in the train, and the requirements for technological procedures at the stations. The transport mass and train length shall be determined in tonnes and meters. The set values must comply with the instructions of SŽDC and ŽSR [11]. If a useful track length is less than the train length norm at the station, this shall be taken into account in the timetable. The transport weight of the train in relation to the regular driving times and consequently the technical norms of weight differ according to the types of driving resistance of individual types of trains. The established driving resistance types (marked as M, R, S, T, and U, for example) are indicated with the train weight normative value. If a freight train path is to be constructed in which the transport weight of the train vehicles is higher than the technical weight standard, then the number, type, and method of deployment of other active traction vehicles (locomotives) must be agreed.

It is important to precisely determine the binding travel time of the train concerned for the construction of the train path.

The methodology for determining the theoretical driving time assumes graphical and numerical solutions to the train equation. It is necessary to construct the tachogram of the train, that is, the algorithm or simulation of the ride, which is tasked with compiling the track and time waveforms by means of the train's differential equations, which result in a graphical dependency of the driving speed on the trajectory $v = f(l)$ as well as dependency of the driving time on the trajectory $f(l)$ [16].

At each stage of the train movement, the traction resistances that are overcome by the tractive force exerted by the driving axles of the tractive vehicle must be taken into account.

Driving time calculation presumptions

Th s section discusses overcoming the traction resistances that occur when starting, running at inertial speed, running, and braking as described in these physical equations [10]:

$$F_t = F_a + F_{0L} + F_{0V} + F_S + F_b \text{ [N]} \tag{1}$$

where F_t is the locomotive pulling force, its graphical representation in relation to speed is traction characteristic $F_t = f(V)$; F_a is the resistance of mass inertia [N]; F_{0L} is the driving resistance of the traction vehicle [N]; F_{0V} is the driving resistance of trailers (loads or wagon set) [N]; F_b is the braking resistance [N]; and F_S is the slope resistance [N].

For the driving resistance coeffici t, which is specifi d for each tractive vehicle and load separately, the empirical relationship applies, where *a, b,* and c are the infrastructure manager's table values:

$$p_0 = \frac{F_0}{G} = \frac{A}{G} + \frac{B}{G} \cdot V + \frac{C}{G} \cdot V^2 = a + b \cdot V + c \cdot V^2 [-] \tag{2}$$

These are the quadratic dependencies and formulas applicable to approved vehicles, as measured by actual measurements on vehicles.

The calculations use the so-called inertial slope, which is defi ed as the slope that is numerically equal to the slope of the line on which a particular train moves at a constant speed. In determining the inertia slope, we proceed from the basic equation of train movement, assuming that the velocity is constant when the acceleration resistance is zero. The graphical dependence of inertia on speed is the so-called $_{s0}/V$ diagram. It is unique to the type of traction vehicle, the type of vehicle resistance, and the weight of the wagon set. Traction characteristics for each traction vehicle (locomotive and motor car) are constructed to obtain a traction force-speed dependence. The vertical y-axis shows the tractive force, and the x-axis shows the speed. The tractive effort curve in the traction diagram indicates a lot about the locomotive's operating characteristics.

The construction of graphical methods to determine the technical normative weight of a train set is based on the theory of nomograms. In practice, SŽDC and ŽSR most often use the Koreff intersection nomogram, constructed under the condition $V = const.$ The tractive force values of the coupler for a given traction vehicle are given by the traction characteristic, and the coeffici t of vehicle resistance can be determined from empirical relationships. The slope of the track is given by the parameters of the track. It follows from the equation of motion (3) that the left side corresponds to the linear dependence on the resistance of the tractive vehicles, and the right side of the linear dependence on the slope. Relations after adjustment for the weight determination of the transported vehicles are represented by two equations of lines whose relationship can be solved graphically.

$$F_{ts} - GD \cdot f_{0V} = (G_L + G_D) \cdot f_S \tag{3}$$

where F_{ts} is the tractive effort of the locomotive on coupler [kN]; G_D is the weight of wagon set [kN]; G_L is the tractive vehicle weight [kN]; f_{0V} is the driving resistance coeffici t of transported vehicles [—]; and f_S is the slope resistance coeffici t [—].

The practical expression of Koreff nomograms for transport practice is tables of the technical normative mass. Tables are compiled for each type of traction vehicle; at the intersection of a certain slope (track class) and the mass of the train set, there is the value of inertial speed that the traction vehicle of the given series is able to haul on a given inertia slope and with a given weight of trailer vehicles. Calculated driving time values are called theoretical driving times, rounded off to at least 0.1 min. Regular driving times rounded to 0.5 min are used for the timetable construction [1, 2, 10].

Identification of problems in railway infrastructure capacity management

For capacity management, the default requirements are:

- organization of the rail market (transport policy and transport market operators);
- technical aspects (infrastructure and interoperability); and
- technological aspects (traffic lanning and management).

On the SŽDC and ŽSR networks, "open access" in freight and passenger transport is also possible. Th s has a major impact on capacity utilization. In the area of railway infrastructure capacity management, major problems in capacity utilization and path allocation have been identifi d. Under the conditions of the SŽDC and ŽSR railway network, new approaches in capacity analysis are defi ed and methodological postulates for new approaches in capacity management in the following areas are defi ed:

- lack of effici t train paths for freight trains on transit corridors;
- application of the integrated tactic timetable for passenger trains; and
- an increase in the demand for "ad hoc" freight paths.

Lack of efficient train paths for freight trains on rail corridors and nodes

The capacity of the lines as a line construction is closely related to the capacity of the individual railway stations, where insuffici t capacity of railway tracks at stations can cause a train to be rejected and wait at intermediate stations, which will further reduce line capacity (and in addition, unproductivity).

For a short case study on the 4th transit corridor, a section of the double-track line between railway stations A and B was chosen, which is characterized by:

- track equipped with a fully automatic track and station security devices, with a short station and track operational intervals as well headways;
- both regional passenger transport and long-distance express and interregional transport are implemented on the line;
- international freight trains and relational freight trains are established on the line; and

- most trains pass through both stations bounding the interstation section under examination without stopping.

The typical timetable for that line is shown in the segment in Figure 1. Due to the preference of passenger transport in the allocation of timetable routes, the assumption can be made that any freight train can only be traced when it does not restrict the movement of passenger trains. The problem is the time taken by a freight train in a timetable, which is signifi antly longer compared to passenger trains.

In the timetable, the driving times for this section are of the order of Ex 4.5–5.0 min, regional passenger trains 11.0 min, freight expresses 7.5 min, and relational freight trains 9.5 min. To do this, there is need to add a start and stop

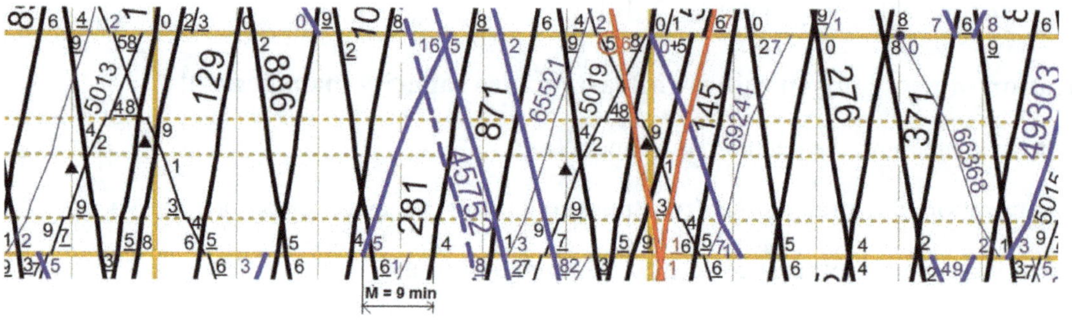

Figure 1.
Headway for train sequence in odd direction (line with odd numbers) stopping freight express-passing express for the station (source: elaborated on ground of [11]).

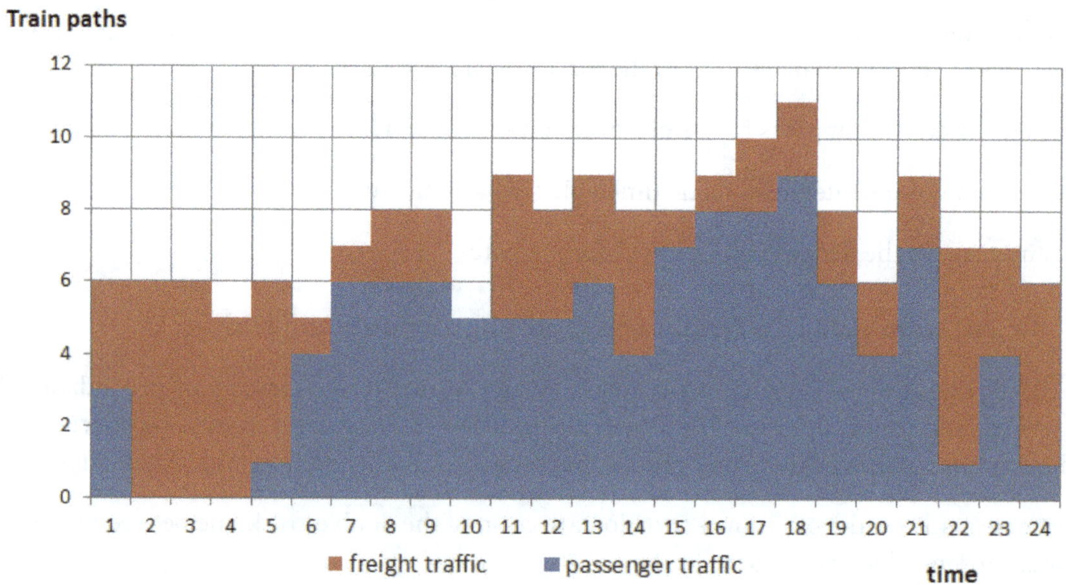

Figure 2.
Histogram of timetable train paths in section A–B.

surcharge of 2.0 min in case of overtaking. From the analysis of time elements, these values of the headway are determined for these train sequences:

- express—express 2.5 min;

- express—freight express 2.0 min; and
- freight express—express 9.0 min.

From these values, it can be seen that the minimum time gap between fast driving trains must be 11.0 min in order to drive the freight express train to the front station where it will be overtaken by the express train. If the freight express train is pathed to the next station, a buffer time of 11.5 min is needed.

The analysis of the constructed timetable in the surveyed section shows that the number of buffer times of at least 11.0 min in this section is on average 1–2 per hour in peak hours. At the same time, the Express-Express trains are most often traced at 8–10 min, which could be reduced to a subsequent interval of 3–5 min (more thorough train bundling).

Regular freight trains are traced a total of 66 trains/24 h, trains as needed five trains/24 h. From the graph in **Figure 2**, which is the histogram of the frequency of passenger and freight trains embedded in individual hours during the day, it is clear that the largest volume of passenger traffi is realized on corridor lines between and 21.00 h. Freight traffi is generated at regular intervals throughout the day, and it can be seen that their journey during peak hours of passenger traffic annot be smooth.

We will assume that freight trains enter the section under investigation at fi ed intervals. It is therefore necessary to examine the time model, refl cting the sequence of freight trains and their ability to travel through the section under investigation, including by using the paths as needed. With 66 trains per 24 h, 2.75 trains are generated every hour and need to be transported. Due to the accuracy of the overall result, the situation is modeled using the distribution of trains processed in **Tables 2** and **3**.

Generalization of the case study conclusions for the capacity and mix of train paths in the timetable of the SŽDC transit corridors:

- the accumulation of passenger traffic etween 6:00 am and 9:00 pm practically stops freight traffi
- at this time, only freight express trains may overrun between express, fast trains, and passenger trains but not slower relational freight trains hauling individual wagon loads or empty wagons;
- time distance between passenger trains in the 9–15 min sequence does not create a suffici t buffer time for the insertion of freight train paths;
- freight trains that cannot pass through the section are waiting at the intersection stations, which mean that on average, four trains must wait 4 h from 4:00 to 8:00 pm;
- rush hour may shift slightly depending on which direction of the passenger traffi is stronger in the morning and afternoon and also depending on the distance of the relevant line from large agglomerations;

Time	**1:00**	**2:00**	**3:00**	**4:00**	**5:00**	**6:00**	**7:00**	**8:00**	**9:00**	**10:00**	**11:00**	**12:00**
Entered freight trains	2.75	2.75	2.75	2.75	2.75	2.75	2.75	2.75	2.75	2.75	2.75	2.75
Stopped freight trains	0	0	0	0	0	0	1.75	3.5	4.25	5	7.75	6.5
Number of	3	6	6	5	5	1	1	2	2	0	4	3

freight trains in timetable												
Overhang of train path requirements over demand	0.25	3.25	3.25	2.25	2.25	-1.75	-3.50	-4.25	-5.00	-7.75	-6.50	-6.25
Number of stopped trains	0	0	0	0	0	1.75	3.50	4.25	5	7.75	6.5	6.25

Table 2.
Average number of delayed freight trains (0.00–12.00 h).

Time	13:00	14:00	15:00	16:00	17:00	18:00	19:00	20:00	21:00	22:00	23:00	24:00
Entered freight trains	2.75	2.75	2.75	2.75	2.75	2.75	2.75	2.75	2.75	2.75	2.75	2.75
Stopped freight trains	6.25	6	4.75	6.5	8.25	9	9.75	3.5	4.25	5	1.75	1.5
Number of freight trains in timetable	3	4	1	1	2	2	2	2	2	6	3	5
Overhang of train path requirements over demand	-6.00	-4.75	-6.50	-8.25	-9.00	-9.75	-10.50	-4.25	-5.00	-1.75	-1.50	0.75
Number of stopped trains	6	4.75	6.5	8.25	9	9.75	10.50	4.25	5	1.75	1.5	0.75

Table 3.
Average number of delayed freight trains (12.00–24.00).

- the dense sequence of passenger trains during the day does not create any room for freight traffi extraordinary balancing. If freight trains are delayed on other sections, for example due to infrastructure, then delayed trains can only be transported after the peak rush hour, usually only at night time with an average stoppage of 10 h;
- exhausting the line capacity with passenger traffi causes measurable economic losses for freight RUs; and
- the free capacity of the track between 11.00 pm and 4.00 am is not used.

These fi dings lead to the need to defi e a powerful train path that can be systematized. The performance of the allocated paths is determined by the RU's data in the capacity allocation request that affects capacity. The analyzed data in Section 2 is mainly about the planned series of the traction vehicle for which regular driving times are computed. It is important to examine the momentum of the train, that is, its mass and acceleration, traction force, and vehicle resistance. These are key parameters that have a major impact on capacity, and the RU can influence these factors. In addition to train dynamics data, the use of infrastructure parameters also affects the level of technical and safety equipment of the traction vehicle. In particular, it concerns the equipment of the traction vehicle with train safety devices and telecommunications equipment, which affect the maximum possible train speed, the use of the transport infrastructure, and the length of its occupation by the train.

To optimize the use of capacity, we have identifi d operational and technological factors affecting the duration of infrastructure use by a train in the process of capacity allocation:

- train speed;
- train performance; and
- equipped with communication and security equipment.

The speed of the train can be distinguished as maximum speed, determined speed, technical speed, sectional speed, etc. As a criterion for assessing the train paths demanded by the RU with the help of the train speed, it is possible to use the relative speed [4]:

$$V_{rel} = \frac{60 \cdot L}{t_{us} \cdot V_{tr}} [-] \quad (4)$$

where V_{rel} is the relative speed; L is length of the examined track section [km]; t_{us} is the travel time of the train on the examined track section including the dwell time; and V_{tr} is the prevailing line speed in the corresponding speed profile in the track section being tracked [km h^{-1}].

Depending on the train's equation of motion (3), the power of the traction vehicle per unit of mass must be evaluated to compare the different train paths. Th s evaluation can be performed according to the criterion of relative power [4]:

$$P_{rel} = \frac{\sum_{x=1}^{i} P_{con}^{x}}{M_{train}} [\mathrm{kWt}^{-1}] \quad (5)$$

where P_{rel} is the relative power [kW t^{-1}]; P_{con} is the continuous power of the traction unit [kW] and i is the number of tractive vehicles; and M_{train} is the train weight [t].

Th s criterion evaluates the actual performance of each examined train path, that is, the momentum of the train (its acceleration) and thus the occupation time of the infrastructure. The above criteria, in particular the criterion of relative speed, are also directly related to the technical equipment of the leading locomotive (tractive unit) by the communication and signaling equipment.

The use of infrastructure capacity, and hence its allocation, must be based on the train's equation of motion and thus on speed and power ratings, using criteria such as relative speed V_{rel} and relative power P_{rel}.

Application of the integrated timetable for passenger trains

Research on the impact of a systematic timetable on infrastructure capacity shows that it reduces track capacity. In the case of the requirements for the fi ed distribution of paths in time, the relevant period is ensured in such a way that at each clock node, it is necessary to reserve a certain part of the capacity before the train runs. Such reserved capacity affects the overall capacity of station heads and track sections. For example, works [17–20], which confi m this, are involved in research in this area. Even the analytical methodologies used to determine the capacity of SŽDC do not affect the construction of systematic interval paths in order to relate the calculation to the average train and to determine the average required buffer time. Th s capacity loss can only be compensated to a certain extent in the case of upgrading of track-side signaling equipment using the ETCS system of application level 3, where only the minimum "moving" track section is reserved for the train departing from the clock node.

Displacement of freight train paths on transit corridors seems problematic. Th s is due to the lack of a sufficient time window to insert the freight train path.

Certain theoretical solutions are offered by the works [21–23]. Freight trains, according to Lindner and von Reder [21], should have periodic time windows between passenger paths, adequate according to the required number of freight routes over the cycle time, into which individual freight train paths can then be constructed. These time windows should, as far as possible, be interconnected between successive line sections (if there is a demand for a freight path from one route to the next one, then the connection of the relevant time window with more follow-up should be sought). In case of insuffici t capacity of the time windows or the necessity to overtake freight trains too often, they propose to review the structure of the passenger transportation offer (e.g., the individual time positions of individual lines or the number of service segments on the given track section). Of course, freight transport requirements must not lead to the breakage of important elements of the network supply, for example (in terms of passenger traffi fl ws) signifi ant connection links.

Drábek [22] in his work presents systematic paths as a network capacity offer, which to some extent is similar to integrated periodic timetable in passenger transport, but takes into account rail freight needs.

These studies demonstrate the difficulty of addressing this issue. Methodologically, a distinction should be made between the procedures for timetable construction with the rail infrastructure adaptation procedures.

In the case of persistent trouble inserting train paths in the required sequence and required dwell times in the timetabling process, there may be a need to change the configur tion of the railway infrastructure. The methodological recommendation for rail infrastructure adaptation procedures only for strategic considerations and objectives can be summarized as follows:

- construction of timetable with a perspective mix of passenger and freight train routes;
- the subsequent defin tion of the necessary infrastructure measures to ensure the implementation of the required scope of transport at the required quality level:
    - to build other station tracks
    - about rebuilding station heads
    - about building rail crossovers
    - about building the next line track
    - about building new safety equipment
- adjusting and synchronizing driving times;
- assessment of the construction of infrastructure (lines and stations) with segregated freight traffic and
- assessment of the construction of high-speed infrastructure with segregated passenger traffi according to the high speed TSI.

Lack of capacity is also due to the lack of useful lengths of station tracks on rail freight corridors (RFCs) [24] as completely unsatisfactory (the study considered suffici t rails with a length of at least 752 m).

The methodological recommendation for the timetabling is intended for intentions in current conditions and the search for technological solutions in the current state of infrastructure, which can be summarized as follows:

- achieving systematization and synchronization of freight train times; this is achieved by correctly determining the driving time, calculated for the selected level of the specifi d speed and the corresponding weight and type of the traction vehicle; to obtain a constructed synchronized path, RUs must meet the performance requirements of this path by providing a train assembly from vehicles that achieve a design speed of at least the specifi d speed and the tractive vehicles have the required performance to ensure the system driving time; and
- achieving systematization and synchronization of passenger train travel times by unifying speed levels for these paths; this means, similarly to freight trains, to create systemized paths with synchronized driving times.

The aim of systemization and synchronization of driving times is to achieve less heterogeneity in train path performance and bundling with minimized buffer time between occupancy times (headways).

A typical example of the distribution of train paths in the timetable on SŽDC resp. ŽSR transit corridors was analyzed in a case study on a corridor line with five

A–E stations on a double track line equipped with an automatic block [10, 16, 25, 26]. The analyzed section can be compared, for example, with the real section Kolín-Pardubice. Passenger trains (Ex category) are run at different times and with different stops. **Figure 3** shows the time spacing between Ex trains approximately 7.5–10.5 min at a headway of 2.5 min. Th s means a request for a buffer time for the path insertion of 5.0–8.0 min. However, the express freight train path needs more time to insert it, namely 12.0 min, which is indicated by the shading of the depicted occupancy time of passenger trains. Th s gray area must not be affected by the occupation time of another train. In this path configur tion, eight freight train paths (freight express and relation freight trains) are inserted in a two-hour time window. The specifi d train speeds in km h^{-1} are listed below the graphical timetable. Ex trains are pathed to 160 km h^{-1}, passenger trains to 140 km h^{-1}, and Nex and relation freight trains up to 100 and 90 km h^{-1}.

Figure 4 shows a study of the distribution of train paths after the systematization and synchronization of passenger train and freight train times. Following the adoption of the methodological recommendation, the Ex train paths are more closely bundled within 5 min to allow for the introduction of freight train paths with a standardized speed of 100 km h^{-1}. In this option, 14 freight train paths were successfully inserted in the 120 min time window. The start-up and stopping time charges for freight trains are problematic and considerably prolong the driving time and affect the possibility of inserting the train path into the buffer time. It would be ideal to achieve the condition that freight trains pass through all stations.

Increase in "ad hoc" train path requirements

From the perspective of the railway infrastructure manager, the growth of requests for "ad hoc" paths, that is, the operationally introduced paths not included in the timetable at the expense of planning regular train paths incorporated in the timetable, is a negative phenomenon.

In the context of individual ad hoc capacity allocation, we divide the capacity requests for "over 3 days," "ad hoc" requests for "under 3 days" capacity, and "ad hoc" allocation for rail capacity for technical-safety tests of railway vehicles and

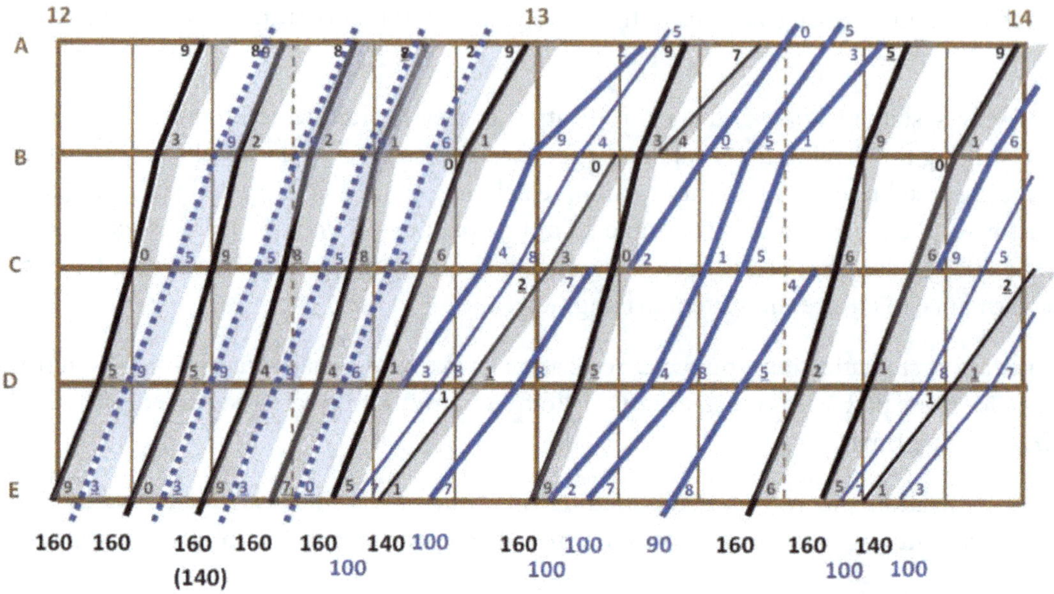

Figure 3.
Inability to insert fast freight train paths (Nex) into gaps between express passenger trains (Ex) indicating the occupation time by those paths in a typical mix of train paths on the transit corridor section (numbers below the paths indicate the specified speed for that path).

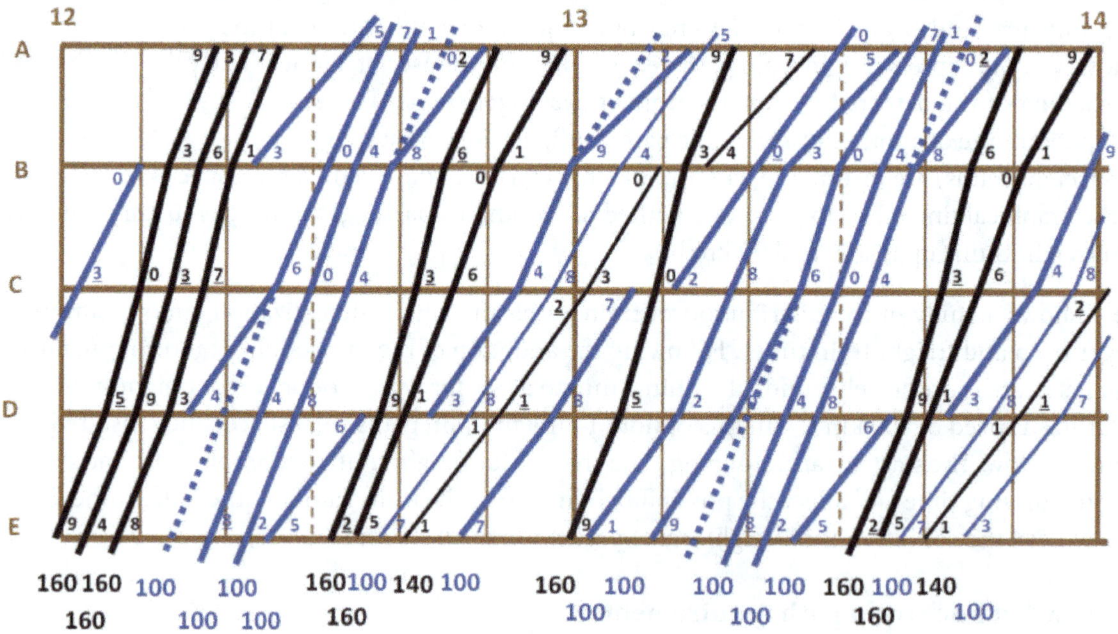

Figure 4.
Example of arrangement of systemized and synchronized train paths for passenger and freight trains (numbers below the paths indicate the specified speed for that path).

other reasons. For "under 3 days" applications, it is up to the infrastructure manager to decide whether to allocate "ad hoc" paths to resolve confli ts (for example, allocate pre-constructed bidding paths) or allocate paths in residual track capacity without confli t resolution. Confli ts in these paths are handled operatively by the operating staff f the rail system operator.

Another administrative constraint is the fundamental difference between the approach of national infrastructure managers to the issue of reservation, allocation, and use of paths. By comparing the conditions on the SŽDC network, the Austrian ÖBB Infrastruktur, and the Polish PKP PLK, it is possible to fi d out that there is no uniform approach in the implementation of partial timetable changes. As a result, there is a situation where the train is already regularly on a single rail network, while on the neighboring rail network it is still in an "ad hoc" mode.

For mutual co-operation between applicants and capacity allocators in the process of allocation of railway capacity, national information systems are used for the setting of the annual timetable, as well as the information system for coordinating the allocated train paths (Path Coordination System) from RailNet Europe.

In the train path request process, it is also necessary to implement the TAF/TAP TSI (Technical Specifi ation for Interoperability relating to Telematics Applications for Freight/Passenger Services) [27] for all rail freight operators in EU Member States. All participants in the transport process will have to be able to exchange precisely defi ed information and reports among themselves electronically. The TAF/TAP TSI will allow coordinating the development of information systems for request acceptance processes, capacity allocation, path design reconciliation, and path activation. However, in the interest of developing rail transport business, infrastructure managers seek to develop these technologies with the least possible fi ancial impact on railway undertakings.

New approaches to capacity management

The basic task of capacity management is to construct a basic timetable for a certain time period (all-year) based on infrastructure capacity planning and specific train path orders. The condition for the allocation of train paths is suffici t

capacity, that is, adherence to defi ed conditions for ensuring the timetable quality, in particular the transportation time, with which the required buffer time (backup time) is closely related. The remaining capacity (free paths) is offered as bidding catalog paths in "ad hoc" mode to RU [1, 6, 11].

The output of the capacity management process is the allocation of train paths and the determination of quantitative and qualitative indicators of the constructed timetable (as a result of the stability proof process), in particular occupancy time, waiting time, buffer time, or optimal traffic w [10, 28].

New approaches to comprehensive management of infrastructure capacity can be broken down into the following headings:

- supporting the implementation of simulation procedures and UIC methodology for capacity utilization;
- marketing measures, in particular capacity management and capacity allocation activities; and
- organizational measures in technology, aimed at systematizing path allocation and operational traffic angement.

The disadvantage for infrastructure managers who use analytical methodologies for capacity determination is that they no longer refl ct the progressive requirements. Th s is mainly due to the development of computer technology and the related possibilities for modifi ation of analytical methods, development of structure, and the heterogeneity of transport (loss of freight transport, development of suburban, and long-distance transport), or the shift from quantitative capacity to qualitative. The use of simulation methods is used to model railway traffi including the inclusion of operational irregularities, that is, delays, to the extent corresponding to reality. An important task is to fulfill the relevant model data, which corresponds as accurately as possible to the reality of infrastructure and vehicle parameters. The modeling of train delays and the feasibility of solving traffi situations have a signifi ant impact on the accuracy of simulation outputs. The outcome of the simulation procedure is to determine the stability of the time- table. Different simulation programs (e.g., RailSys, OpenTrack, and SimuT) provide different results (using different ways of calculating driving times, solving confli ts between trains, etc.) [14]. Investigating the reduction of the initial delay means determining the average delay increase per train. An increase in delay of up to 0.5 min/train may still be acceptable, but this increase should be able to absorb adjacent infrastructure elements. It is recommended to increase the occupancy degree limit in the following cases [16]:

- in a peak computing period;
- average occupancy time is greater than 10 min; and
- it is a track with a specific traffi (e.g., only one type of trains prevails on a track that achieves low delays).

These indicators can be fully explored in simulation procedures supported by UIC Regulation 406 "Capacity." Principal differences in analytical and simulation approaches in capacity exploration are shown in **Table 4**. There is no exact dependency between the degree of occupation and the quality of traffic so analytical methods are less accurate [14, 16].

Methodology	Analytical methods (SŽDC, ŽSR)	Simulation methods (UIC)
Quality of operation	Based on the degree of occupancy of individual devices	Using quantities directly describing the quality of operation (delays)
Capacity relationship	Capacity related to infrastructure capacity	Related to specific timetable
Deterministic	Based on even track occupancy; unequal track occupancy can be used for tracks used only by freight transport	Uneven track occupancy

Source: [10].

Table 4.
Basic differences in capacity research approaches.

Marketing measures are aimed at achieving optimum use of infrastructure (more evenly burdened) by using pricing and other tools in the capacity allocation process, in particular:

- allocation of train path depending on its time position;

- taking into account the acceptance of powerful train paths;
- deviations from the time position, that is, penalty for delay caused by the carrier, as well as delay bonuses caused by the infrastructure manager (EPR system);
- sanctions for nonuse of allocated capacity;
- stricter conditions for "ad hoc" capacity allocation; and
- the scope of use of ancillary services.

The essence of organizational measures is the systematization and synchronization of driving times, where the aim is to achieve less heterogeneity of train paths and to bundle it with minimizing buffer times. There is a need for a change in the configur tion of the railway infrastructure in the event of persistent problems with the insertion of train paths in the required sequence and the required timetable stays. In the operative operation, the fi ed specifi ation and order of the train paths will be followed. In the case of delay, the right to the train path will not be guaranteed.

Capacity management processes include the train path ordering processes, strategic dialog and advisory phases, confli t coordination and resolution, and capacity allocation. The aim is to create a comprehensive, effective, and motivating model, which will be composed of both technological procedures and pricing policies leading to the provision of effici t train paths.

A prerequisite for creating a capacity management concept is capacity planning for its further development and use. The essence of capacity management is based on a plan to create a basic timetable, which is usually 1 year and thus represents a medium-term plan. The output of the capacity management process is the allocation of train paths and the determination of the quantitative and qualitative indicators of the constructed timetable (as a result of the stability proof process), in particular the occupancy time, waiting time, standby time, and optimum traffi fl w. Achieving optimum traffi fl w over time requires infrastructure managers to use capacity management and marketing tools to allocate train paths to refl ct traffi fl w over time so as to get as close as possible to the defi ed optimal value. The fi st priority is the technical conditions for access to infrastructure and the second is the infrastructure access clearing system [2, 10, 16, 25, 28–31].

The aim is to create a comprehensive, effective, and motivating model consisting of both technological procedures and pricing policies, leading to the provision of effici t train paths:

- allocation of the train path depending on its time position, that is, providing more favorable conditions for the allocation of the path in the transport saddle;
- taking into account the acceptance by the carrier of effici t train paths;
- deviations from the time path, that is, the penalty for delays caused by railway undertakings, as well as delay bonuses caused by the Infrastructure Manager (European Performance Regime);
- when using capacity utilization, to reconcile the use of standby time and occupancy with the UIC methodology, in particular to defi e the upper occupancy level on lines with specific traffi (homogeneous timetable), which can be as high as 0.90;
- sanctions for nonuse of allocated capacity; and
- stricter conditions for ad-hoc capacity allocation.

Figure 5 shows the identifi d cycle of capacity allocation of the railway infrastructure using the proposed progressive methodological approaches. It is a set of methodological postulates in the following defi ed areas:

- setting a higher level of optimal capacity utilization in a specific imetable;
- determining the prioritization of train paths in the construction of the timetable, as well as in the operative management of traffic ith the help of defini g system times; and
- connecting the required heterogeneity and sequence of train paths, to provide the required performance paths, and to propose infrastructure measures if the confli ts cannot be resolved by technological measures.

These methodological proposals are only a simplifi d procedure for timetable creators. If they are to be the basis for further processing, especially by means of IT techniques, it is necessary to create a mathematical model that will be revised to a computer model after appropriate verifi ation. After validation, it is the basis for the creation of an applied computer program. The proposed methodologies can then be imported into software products supporting the construction of the order and capacity indicators.

In the liberalized railway market, further development of infrastructure capacity utilization in:

- searching for optimum traffic w in order to achieve the desired quality of transport;

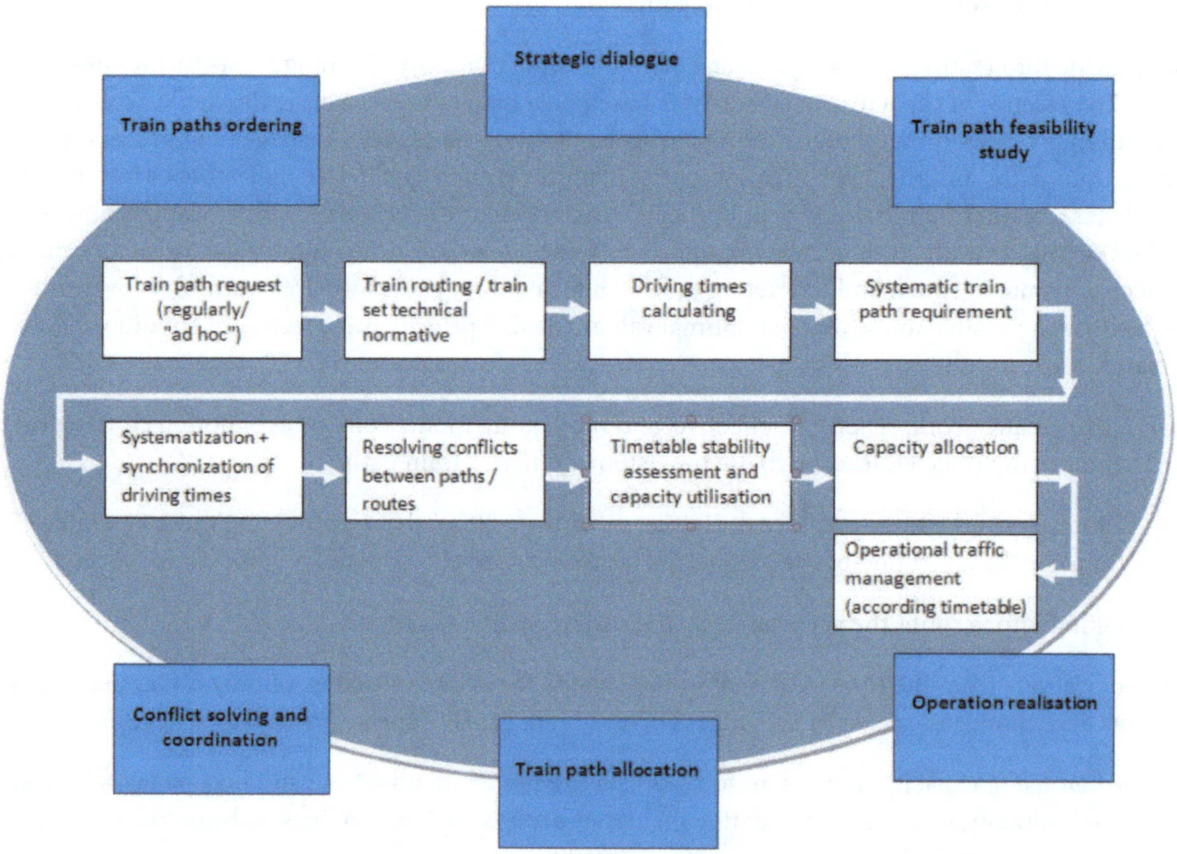

Figure 5.
Capacity allocation cycle using proposed progressive methodological approaches with regard to the railway infrastructure capacity.

- establishing a timetable for each day (planning period of 6 or 12 h), where operational planning and the interconnection of tools for timetabling and capacity planning with operational planning systems will be supported;
- expressing capacity as the ability to meet the requirements of specific train paths while guaranteeing the stability of the timetable operational model; and
- planning infrastructure configur tion based on required capacity through infrastructure measures.

Conclusion

Sizing the railway infrastructure, operational performance and quality of operation are interrelated. If two variables are known, the third can be derived. For example, on the Prague-Ostrava route, the current trend of shortening the length of long-distance passenger trains but increasing their frequency as a result of the liberalization of the rail market ("open access") signifi antly affects the capacity. From this situation, there is a clear need to plan a transport infrastructure for the future scope and concept of transport. Emergency events are a problematic point in the framework of the liberalization of rail transport, which is related to maintaining the number of traction units to the minimum possible. As a rule, RU does not have suitable tractive vehicles for diversion routes. The former unitary state railways had tractive vehicles for independent traction (multi-system locomotives), which could be used for such haulages (not only for freight).

In order to create harmonized conditions for access to infrastructure, it is important to correctly determine the optimum capacity of the railway infrastructure, that is, the number of train paths that will be advantageous both for the infrastructure manager and operating and economical. The prerequisite for this is a clear defin tion of the capacity of railway infrastructure and train paths, determination of the basic principles of capacity detection by means of simulation tools, determination of optimum throughput, plus even ensuring the utilization of railway infrastructure capacity. The proposal outlined the process of determining the capacity of the railway infrastructure, including progressive capacity allocation approaches as key components of capacity management. The scientific and research contribution is proven in the focus, deepening the application of theoretical knowledge in the fi ld of railway transport technology. The proposed procedures for determining the timetable stability and optimizing the process of inserting train paths into the timetable, with an emphasis on optimizing and systematizing them, can be applied as a manual for the needs of specific simulation procedures, as well as for the support software product developers. These benefits can be used in the practical tasks of allocating railway infrastructure capacity and identifying optimal traffi fl w. At the same time, the proposed procedures for determining graphical stability can be applied as a manual for the needs of specific simulation procedures, as well as for creators of supporting software products.

The issue being solved by its impact on traffi and transport operations and on the railway market is so complex that a number of related aspects are subject to further research. When establishing capacity, focus should be on the required amount of buffer time and examining the waiting time in terms of overall capacity management and in the context of streamlining and shortening the entire process of allocating railway infrastructure capacity. These tasks have a signifi ant impact on the main objective of European rail transport policy in order to strengthen its competitiveness in sustainable development.

Acknowledgements

The chapter is supported by the VEGA Agency by the Project 1/0509/19 "Optimizing the use of railway infrastructure with support of modal split forecasting," which is solved at Faculty of Operations and Economics of Transport and Communication, University of Žilina.

Author details

Jozef Gašparík[1]* and Václav Cempírek[2]

Faculty of Operation and Economics of Transport and Communications, University of Žilina, Slovakia

College of Logistics, Přerov, Czech Republic

*Address all correspondence to: jozef.gasparik@fpedas.uniza.sk

References

[1] Gašparík J, Šulko P. Technológia železničnej dopravy—líniové dopravné procesy (Railway Transport Technology — Line Transport Processes). 1st ed. Žilina: University of Žilina; 2016. p. 383

[2] Široký J, Gašparík J, Abramović B, Nachtigall P. Transport Technology and Traffi Management. 1st ed. Pardubice: University of Pardubice; 2018. p. 254. ISBN 978-80-7560-163-6

[3] Dedík M, Gašparík J, Záhumenská Z, Ľupták V, Hřebíček Z. Proposal of the measures to increase the competitiveness of rail freight transport in the EU. Naše More (Our Sea). 2018; 64:202-207

[4] Štěpán O, Kopecká P, Molková T. Czech Hodnocení a Kvantifikace vlastností Tras vlaků při sestavě jízdního řádu (Evaluation and Quantifi ation of Train Path Characteristics for the Timetable) Horizons of Railway Transport. Strečno, Žilina: University of Žilina; 2015

[5] Network Statements. RailNet Europe. Available from: http://www.rne .eu/organisation/network-statements/

[6] Dolinayová A, Černá L, Zitrický V. The role of railway transport in eastwest traffi fl w conditions of the Slovak Republic. In: Sladkowski A, editor. Transport Systems and Delivery of Cargo on East-West Routes. 1st ed. Cham: Springer International Publishing AG. pp. 121-169. DOI: 10.1007/978-3-319-78295-9

[7] Siroky J, Salakova H. Analysis of the system of pricing of railway capacity allocation in the Czech Republic. In: Horizons of Railway Transport 2018, MATEC 235 (2018) 00016. ISSN 2261-236X. London: EDP Sciences; 2018, 2018. DOI: 10.1051/matecconf/201823500016

[8] Malavasi G, Ricci S, Rizzetto L, Rotoli F, Valeri E. Potential effects of changes in the Italian railway access charge regime on passenger services. Ingegneria Ferroviaria. 2019;73(3): 183-206

[9] Handbook for the European Performance Regime (EPR). Guidelines for Actual and Potential Users. 1st ed. Joint Project by UIC/RNE; 2013

[10] Gašparík J, Kolář J. Železniční doprava—Technologie, Řízení, Grafikony a Dalších 100 Zajímavostí (Railway Transport—Technology, Management, Timetabling and 100 Other Attractions). 1st ed. Prague: Grada Publishing; 2017. p. 432

[11] Internal Rules of Infrastructure Managers SŽDC, ŽSR, PKP PLK

[12] Leaflet UIC 406. Capacity. Paris: International Union of Railways

[13] Kontaxi E, Ricci S. Techniques and methodologies for carrying capacity evaluation: Comparative analysis and integration perspectives. Ingegneria Ferroviaria. 2009;64(12):1051-1080

[14] Gašparík J, Abramović B, Zitrický V. Research on dependences of railway infrastructure capacity. Tehnički Vjesnik. 2018;25(4):1190-1195

[15] Directive No. 2012/34/EU of the European Parliament and of the Council of 21 November 2012, Establishing a Single European Railway Area

[16] Kolář J. Progressive approaches in the management of railway infrastructure capacity (in Czech) [thesis]. Prague: Czech Technical University; 2018

[17] Lichtenegger M. Integrated Timetable, Mapping and Construction with the Help of Graph Theory, Minimization of Realization Costs (in German). Graz: Technical University; 1990

[18] Stohler W. Integrale Taktfahrpläne und S-Bahnen (Systematic timetables and S-Lines). ETR. Vol. 461997. pp. H1-2, 33-38

[19] Nachtigall K. Periodic Network and Optimization and Fixed Interval Timetables. Braunschweig: Deutsches Centrum für Luftund Raumfahrt e.V; 1999

[20] Hrabáček J. Periodická doprava na dopravních sítích a její optimalizace (Systematic traffi on the transport networks and its optimalisation) [thesis]. University of Pardubice; 2010

[21] Lindner HR, von Redern HW. Güterzüge im Taktfahrplan— Möglichkeiten und Grenzen (Freight trains in systematic timetable— posssibilities and limitations). Die Bundesbahn. 1989;10:867-874

[22] Drábek M. Periodic freight train paths in network [thesis]. Prague: Czech Technical University; 2014

[23] Janoš V, Drábek M, Michl Z. Quantitative determination of bottlenecks in railway networks with periodic service. In: 20th International Conference Transport Means 2016; 5–7 October 2016; Juodkrante. Kaunas: Kauno Technologijos Universitetas; 2016. pp. 594-598. ISSN 1822-296X

[24] Regulation (EU) No 913/2010 of the European Parliament and of the Council of 22 September 2010 Concerning a European Rail Network for Competitive Freight

[25] Zitrický V, Černá L, Abramović A. The proposal for the allocation of capacity for international railway transport. Procedia Engineering. 2017; 192:994-999

[26] Vojtek M et al. The purpose of longdistance passenger trains in public passenger transport system. Transport and Communication. 2017;2:37-40

[27] Directive 2008/57/EC of the European Parliament and of the Council of 17 June 2008 on the Interoperability of the Rail System within the Community

[28] Hansen IA, Pachl J. Railway Timetabling & Operations. 1st ed. Hamburg: Eurailpress; 2014

[29] Stopka O, Ponicky J, Chovancova M, Zitricky V. Draft method for determining the number of checking devices utilized within the regional passenger transport. Naše More: Journal of Marine Science and Technology. 2016;63:200-203. DOI: 10.17818/NM/2016/SI23

[30] Ľupták V, Droździel P, Stopka O, Stopková M, Rybicka I. Approach methodology for comprehensive assessing the public passenger transport timetable performances at a regional scale. Sustainability. 2019;11:3532. DOI: 10.3390/su11133532

[31] Sipus D, Abramovic B. The possibility of using public transport in rural area. In: 12th International Scientific Conference of Young Scientists on Sustainable, Modern and Safe Transport. TRANSCOM 2017— Procedia Engineering. Vol. 192. 2017. pp. 788-793. DOI: 10.1016/j. proeng.2017.06.136

10

Third-Party Logistics

Yangyan Shi, Rafay Waseem and Hafiz Muhammad Shahid

Abstract

Third-party logistics was propelled first in US; later on, European nations put it on to successfully deal with an organization's coordination exercises, re-appropriating the capacity and purchasing the administrations. 3PL suppliers without their very own advantages are called lead coordination suppliers that have preferred the standpoint that they have particular industry ability combined with low overhead expenses, yet they bring down arranging power. Lead coordination suppliers may likewise be less bureaucratic with shorter basic leadership cycles because of the littler size of the organization, and the most critical contrast between a second gathering coordination supplier and an outsider coordination supplier is the way that a 3PL supplier is constantly incorporated in the client's framework. The 2PL is not coordinated as compared to the 3PL as it is just a redistributed coordination supplier with no framework mix. A 2PL regularly just gives institutionalized administrations, while 3PLs frequently give benefits that are redone and particular to the necessities of their client. Coordination is evaluated as a significant use for organizations. Thus, in the present aggressive condition, there is a squeezing need to control coordination expenses, and execution estimation has turned out to be an effective apparatus in accomplishing business targets.

Keywords: third-party logistics, types of 3PL, layers, information technology, logistics

Introduction

Globalization has fundamentally changed the business environment. In response to this new economic reality, manufacturers and suppliers have embraced third- party logistics (3PL) providers as key players in the supply chain [131]. As dedicated logistic specialists, these fi ms allow their customers to concentrate on their core competencies and this focus can become a source of competitive advantages. According to [15], coordination redistributing is characterized as common understandings among assembling and administration fi ms with 3PL suppliers. Lieb [75] defi ed outsider coordination as redistributing coordination works recently actualized in-house, through outer organizations. Subsequently, outsider coordination can be divided as dealing with numerous coordination benefits by outer redistributing office in lieu of a customer global organization. Outsider coordination benefits for the most part concentrate on transportation, warehousing, etc., and these 3PL specialist co-ops ought to have proficie t involvement in each administration [23]. Due to the expanding signifi ance of coordination re-appropriating, choosing the right 3PL is a basic issue among organizations. Amid the most recent decades, most investigations directed on 3PL in the writing predominantly identify with papers dependent on MCDM, measurable techniques, computerized reasoning, scientific programming, and coordinated strategies [35]. Marasco [82] provided a writing audit on 3PL characterizing 152 articles distributed somewhere between 1989 and 2006 dependent on substance and technique. Aguezzoul [4] analyzed around 67 articles between 1994 and 2013 dependent on 3PL criteria and strategies in different MCDM systems, numerical programming, and measurable methodologies. Among them, half-breed techniques are used to class the most critical evaluation criteria and to choose the best 3PL supplier [108].

Further, no similar examination of the 3PL suppliers is performed utilizing ordinal, limited, and stochastic information [77]. The consistency in their exhibited structure is not verifi d and the mind-boggling choice emotionally supportive network (DSS) created is not extremely easy to use for the production network directors [63]. Recognizably, their proposed strategy just gives target data to 3PRLP determination without permitting space for the emotional etymological data. Chen and Wu [23] consolidated Delphi and expository system process (ANP) approach to survey coordination benefit providers in the gadgets division. Be that as it may, there is just slight variety in correlation result with their proposed model with existing AHP technique taking variety in estimations of 3PL criteria, subsequently neglecting to demonstrate the heartiness and preferred standpoint of their recommended methodology. Zhang et al. [144] proposed a scientific model consolidating data granulation entropy approach, K-implies bunching, and TOPSIS strategy for choosing a 3PL supplier. Here their investigation neglects to think about the precision of the proposed IGET (data granulation entropy-based TOPSIS) approach in 3PLs, with other existing strategies due to non-presence of benchmark esteems. Likewise, the IGET strategy neglects to plainly look at the signifi ance of different criteria [37]. Incorporated strategy is suggested strategy that joins AHP, DEA, and direct programming (LP) to assess and select the best 3PL supplier in Italy.

Logistics

In the present business setting, China depends more on local utilization for fi ancial development, which is probably going to additionally animate the interest for coordination and transportation benefits in the nation [49]. Since the presentation of outsider coordination (3PL) benefits in the mid-1990s and especially after China joined the World Exchange Association (WTO) in late 2001, the 3PL business has grown quickly [26, 27, 109]. The aggregate estimation of interest for coordination expanded from RMB 59.6 trillion (US$ 9.6 trillion) in 2006 to roughly RMB 120 trillion in 2016 (US$ 20 trillion) (Exploration and Markets, 2017), speaking to a normal yearly development of 9.5% amid this period. Ongoing development of the Chinese internet business industry [57, 58] evaluated that the showcase size of online business in 2019 will be US$1.97 trillion (advertiser, 2016) and is relied on the development of the Chinese 3PL industry [57, 58].

A later report by [79] fi ds that hypothetically grounded 3PL research in China is constrained. Besides, in studies where hypothesis is connected, it depends on a sole hypothesis to clarify a specific marvel. Given the logical multifaceted nature in China, utilization of different hypothetical establishing is useful to pick up a more extensive comprehension of the exercises of MN3PLs [79]. We consider three hypotheses, for example, TCE, RBT, and neo-institutional hypothesis (NIT) as hypothetical underpinnings of our exploration. Th ough a broad writing audit, this examination recognizes 14 difficulties for MN3PLs and gathers them into four test classes, for example, money related, data innovation, HR, and business scene.

Exchange cost fi ancial aspects (TCE): TCE gives a solid establishment to break down coordination redistributing choices [8, 119] which expresses that a fi m composes its authoritative exercises to limit its generation costs inside the fi m and limits exchange costs inside the market [135]. Here the fundamental standard is that re-appropriating exercises to 3PL specialist organizations will happen when there is a chance to decrease exchange costs. Studies propose that re-appropriating coordination works ordinarily lessens exchange costs that incorporate request combination, unifi d request preparing, proficie t utilization of advantages, and solidifi ation of overhead by an outsider [97, 143]. It is patent that the connection between a redistributing fi m (purchaser) and a 3PL specialist co-op (vender) will acquire exchange costs; nonetheless, co-activity, cooperation, and the opportune sharing of data among fi ms (purchaser dealer) in an inventory network relationship can decrease exchange costs [20, 51].

Regularly, 3PL specialist organizations work in a business situation where they continually experience administrative weights. Th ough a broad writing audit of both Western and Chinese 3PL research, we recognize 14 difficulties for MN3PL specialist co-ops working in China. These difficulties are then assembled into four larger amount test classifi ations, for example, money related, mechanical, human asset, and business scene. Money-related difficulties considered in this investigation are value weight, high transport cost, and budgetary dependability. Being an asset, budgetary capacity can give an upper hand and is one of the vital factors in building between fi m organizations and vital collusions [64]. Every one of these difficulties is talked about in the following sections.

Since 3PL clients are value delicate, they always put a solid accentuation on cost while assessing the coordination benefit re-appropriating choices [126]. To prevail in such a value delicate condition, 3PL suppliers need to enhance their capacities and offer administrations at lower costs [117, 143]. The test is that 3PL suppliers should be cost cognizant to accomplish higher edges and to keep themselves from being valued out of the market because of expanded nearby work costs, fuel costs, protection premiums, and rising client benefit desires [59]. In China, transportation costs covers normally 54.2% aggregate coordination cost (KPMG, 2016). The Chinese government has invested extensively in the coordination departments to enhance dissemination frameworks [42]. Be that as it may, insuffici t coordination framework, high rate of harm in travel, clog at ports and modern streets, and confli ting strategies and directions include extra weight transportation [143, 144].

Money-related strength of 3PL is placed as one of the basic perspectives of their tasks and one of the noteworthy criteria for picking a 3PL fi m [59, 143]. A monetarily solid 3PL supplier has assets to contribute for clients and can improve its own operational capacities [121] and fulfill its customers, diminish hazard for coordination accomplices, and impact the long haul connections [97]. IT limit and ability are 3PL fi ms' inside assets that can give a continued upper hand [17]. Writing proposes that IT abilities recognize MN3PLs from the other three kinds of nearby 3PL fi ms in China [124]. Th ough IT frameworks (e.g., mechanized process), a 3PL supplier can create powerful correspondence with its clients. Besides, IT foundation bolsters synergistic interorganizational connections by diminishing exchange expenses and dangers related with robotized forms [29]. For MN3PL suppliers in China, IT challenges considered in this investigation are identifi d with capacity, security, similarity, and unwavering quality of data innovation [139]. A short portrayal of every one of these difficulties is given below.

IT ability of 3PL suppliers is a standout among the most basic assets influencing the choice of fi ms to re-appropriate to 3PL suppliers [100]. To secure IT capacities, fi ms need to spend assets [78, 136]. The generally speaking test to 3PLs in regard to IT ability is to guarantee that staff has the learning and aptitudes, and comprehension to convey the required dimension of IT usefulness and expectations. In the meantime, it is important to guarantee that the PC and interchange equipment and programming are suitable to include huge esteem and enhance execution for the inventory network accomplices [122]. Security and secrecy of data are imperative issues and can specifi ally influence 3PL specialist co-ops' trust improvement [10–12, 96]. The two shippers and clients trust 3PL specialist co-ops to keep their secret information flawless and blocked off to outsiders under all conditions [98, 99]. The sharing of data among included gatherings defin tely prompts security concerns; the data in connection to every one of them ought to be kept private.

The proposed MN3PL challenge system comprises of an organized chain of important test classes and diffi ulties. To recognize the basic difficulties of MN3PL, in the proposed structure, chiefs need to allocate abstract need weights for judgments.

Lead logistics providers

3PL suppliers without their very own benefits are called lead coordination suppliers. Lead coordination suppliers have the preferred standpoint that they have particular industry skills combined with low overhead expenses, yet they bring down arranging power and less assets than an outsider supplier who depends on a regularly huge organization measure, a great client base, and built-up system frameworks. 3PL suppliers may forfeit proficie cy by inclining toward their very own benefits with the end goal to augment their own effectiveness. Lead coordination suppliers may likewise be less bureaucratic with shorter basic leadership cycles because of the littler size of the organization.

Yet, the evaluation of 3PL specialists signifies the subjective criteria for tire fabricating industry [55]. Additionally, affectability examination for checking the overall signifi ance of 3PL administration criteria does not appear in this methodology [141]. Here, the changes in the quantity of potential 3PL specialists or the assessment criteria hardens the procedure [103]. In spite of the fact that the idea of IVFNs is investigated here by the creators to enhance criteria assessment appraisement file framework, computer-based strategy is the need in this examination.

Information technology (IT)

The IT frameworks of the supplier and the customer must be interoperable. Innovation is the best to perceive ability for the customer by method for consistent notices by means of Dispatch Administration Programming and Electronic Information Exchange (EDI) which involves an expense, however it can help maintain a strategic distance from punishments for postponements and ensuing money related misfortunes, for example, from not emptying cargo in time [47].

Evaluated the interrelationships between dangers looked by 3PL suppliers in between classification connection to their partners utilizing DEMATEL. Be that as it may, their exploration needs generalizability by taking just two 3PLs in the specific situation. The creators likewise neglect to catch and orchestrate contributions to uncertain variable, making it risky for specialists in giving just fresh data sources. Keshavarz Ghorabaee [66] proposed another incorporated methodology dependent on the criteria significance between criteria connection (Commentator) and Weighted Accumulated Total Item Appraisal (WATIA) strategies to assess 3PL suppliers with Interim Kind 2 Fluffy sets (IK2FS). In perspective of reasonable uncertain 3PL criteria weights, both emotional data communicated by leaders and target information of choice framework are to be considered in the assessment procedure. Jung [61] addressed 3PL supplier assessment issue considering social maintainability and applying fluffy AHP in assessment structure. Here, the examination neglects to consider analysts' pre-assessment in checking the possibility of the 3PL criteria, and natural measurement was likewise not considered here. Ecer [35] combined fluffy AHP and assessment dependent on separation from normal arrangement (EDAS) for the determination of a legitimate 3PL supplier. It may seem the determination show, we see that expansion in number of criteria makes intricacy in the basic leadership by heightening the likelihood of holes in emotional judgments. Consequently, a writing study based on existing strategies with their examination holes is quickly talked about. Be that as it may, no adjustment of the BWM alongside WASPAS multi-criteria basic leadership procedures by applying interim harsh number (IRNs) has been broken down in the writing up until this point. As of late, IRN-DEMATEL [92], IRNMAIRCA [92], and IRN MABAC [93] are created.

Coordination is one of the dynamic exercises that empowers the association among generation and utilization [14]. As per the Gathering of Production network Administration Experts, coordination comprises of an arrangement of procedures enveloping arranging, executing, and controlling the stream of merchandise, benefits, and related data [127, 128]. Coordination is a mind-boggling business and can be estimated from alternate points of view. One of the goals of coordination is to ensure the effectiveness and the adequacy of the considerable number of methods from the purpose of beginning to the point of goal while meeting the clients' required quality, including data dependability and sensibility to clients' needs. Coordination is not applicable for the creation division; however, it is additionally vital for endeavors from all portions, e.g., banks, retailers, government, and foundations. Coordination assumes a key job in the aggressiveness of associations while making an incentive by giving time and place utility. Waters [131] alludes that "Without coordination, no materials move, no activities should be possible, no items are conveyed, and no clients are served." To position the correct items near the correct buyer, a few exercises must be performed, including transport, client benefit, data innovation and interchanges, back, warehousing, and re-appropriating [41]. With the end goal to play out these exercises, the investment of a few performing artists is required: cargo forwarders, transporters, outsider coordination suppliers (3PL), distribution centers, shipping organizations, producers, and retailers, to list a few.

Layers

First gathering coordination suppliers (1PL) are single specialist organizations in a particular geographic zone that have some expertise in specific merchandise or dispatching strategies. Models are conveying organizations, port administrators, and station organizations. The coordination bureau of a delivering fi m can likewise be a fi st gathering coordination supplier on the off chance that it possesses transport resources and warehouses [90].

Second gathering coordination suppliers (2PL) are specialist co-ops that give their specific coordination benefits in a bigger (national) land zone than the 1PL do. Regularly, there are outline contracts between the 2PL and the client, which control the conditions for the vehicle obligations that are for the most part put here and now. 2PL's coordination assets like trucks, forklifts, and distribution centers are ready for transport, treatment of load or stockroom administration activities [90]. Second-party coordination emerged over the span of the globalization and the uprising pattern of lean administration, when the organizations started to re-appropriate their coordination exercises to center around their own center organizations. Precedents are messengers, express and bundle administrations; sea bearers, cargo forwarders, and transshipment suppliers.

The most noteworthy distinction between a second gathering coordination supplier and an outsider coordination supplier is the way that a 3PL supplier is constantly incorporated in the client's framework. The 2PL is not incorporated, as opposed to the 3PL he is just a re-appropriated coordination supplier with no framework combination. A 2PL is frequently accessible if the need arises (e.g., express package administrations) though a 3PL is pretty much educated every time about the outstanding task at hand in the near future. The difference between 2PL and 3PL is the particularities and modifi ation of administrations. A 2PL regularly just gives institutionalized administrations, while 3PLs frequently give benefits that are tweaked and concentrated to the necessities of their client. Cost viability of an outsider coordination supplier is just given over extensive stretches of time with stable contract and benefits. As opposed to that, second gathering calculated administrations cannot be modifi d, due to the fluctuating business sector with hard rivalry and a value fi ht on a low dimension. Furthermore, there we have another distinctive point somewhere in the range of 2PL and 3PL: Strength of agreements. 3PL contracts are long haul contracts, while 2PL contracts are of a low sturdiness, so the client is adaptable in reacting to market and value changes. With organizations working comprehensively, the need to build production network deceivability and decrease chance, enhance speed, and diminish costs—all in the meantime—requires a typical mechanical solution [2]. Non-resource–based suppliers perform capacities, for example, counsel on bundling and transportation, cargo citing, money-related repayment, reviewing, following, client administration, and issue resolution [1]. Be that as it may, they do not utilize any truck drivers or stockroom staff, and they do not claim any physical cargo conveyance resources of their own—no trucks, no capacity trailers, no beds, and no ware-housing. A non-resource–based supplier comprises a group of area specialists with aggregated cargo industry aptitude and data innovation resources. They fill a job like cargo operators or representatives, yet keep up an altogether more prominent level of "hands on" inclusion in the transportation of items. These suppliers are 4PL and 5PL administrations.

A fourth gathering coordination supplier has no claimed transport resources or distribution center limit. They have an allocative and mix work inside an inventory network with the point of expanding its proficie - cy. The possibility of a fourth-party coordination supplier was conceived in the 1970s by the counseling organization Accenture. Firms are re-appropriating their determination of outsider coordination supplier and the enhancement procedure of the reconciliation of these to a PL as a mediator. That decreases costs and the 4PL must have a diagram about the entire coordination market to pick the perfect 3PL for all agent-calculated exercises. For having the capacity to give such a perfect arrangement, fourth gathering coordination suppliers require a decent learning of the coordination branch and a decent IT foundation. A fourth gathering coordination supplier chooses the 3PL suppliers from the market that are most appropriate for the strategic issues of his client. The capacity of a 4PL in the production network is the main feature of a 3PL supplier as the agent logistics [60].

Fifth gathering coordination suppliers (5PL) give production network administration and offer framework situated counseling and inventory network administrations to their clients. Progressions in innovation and the related increments in inventory network deceivability and between organizations' correspondences have offered ascend to a generally new model for outsider coordination activities—the "non-resource–based coordination provider" [40].

Types of third-party logistics

Thi d-party logistics suppliers incorporate cargo forwarders and dispatch organizations and also provide different organizations' coordinations, offering subcontracted coordination and transportation administrations. Susanne and Monica [50] describe four classifi ations of 3PL suppliers.

Standard 3PL supplier

Th s is the most fundamental type of a 3PL supplier. They would perform exercises, for example, pick and pack, warehousing, and conveyance (business)— the most fundamental elements of coordination. For a dominant part of these organizations, the 3PL capacity is not their primary action. Globalization has in a general sense changed the business condition. In light of this new money-related reality, creators and suppliers have gotten a handle on third social gathering collaborations (3PL) providers as key players in the generation organization. As submitted key stars, these associations empower their customers to center around their inside capacities [112, 143] and this middle can transform into a wellspring of high grounds [6–19, 54]. The dependence that re-appropriating makes can test customer-supplier relations [70, 89, 130]. Fundamental work observes that notwithstanding the way that resistance is a typical for any free market, both trust and duty are vital to keep up a productive exchange. In these incredible and dependent associations, accomplishment or dissatisfaction relies upon the proximity of cooperation and the nonattendance of shrewd practices. Enduring composing has found that trust can propel joint effort and information sharing between dependent associates, which convert into extended operational effici cy [34–43, 91].

Administration designer

Th s sort of 3PL supplier will offer their clients propelled esteem included administrations, for example, following and following, cross-docking, particular bundling, or giving an extraordinary security framework. A strong IT establishment and an attention on economies of scale and extension will empower this sort of 3PL supplier to play out these kinds of undertakings. The dynamic limit suggests fi ms that work in powerful markets need to gain, facilitate, reconfigu e, and release advantages to make a whole deal high ground [46, 76]. These dynamic limits introduce plans that affiliate to make regard and respond to publicize powers. Firms create both convincing and helpful plans that are affected [36, 129]. In spite of the way that analysts have considered the activity of learning presentation on forceful execution [11–25, 38, 44, 105, 123, 140].

Its impact in the key setting says, all things considered, are unexplored [21]. Panayides [94] provided a structure on how various leveled learning adds to compelling relations between collaborations, expert communities, and their customers, yet additional observational examination that extends this model is advocated. Learning can give a viable high ground to fi ms, especially in a sketchy and forceful condition. In any case, learning is not an inherent typical for a fi m; it is the consequence of a culture that hopes to make it. Learning acquaintance is a fondness with development the anchoring of data and fi ms that regard it will as a rule ask delegates to address various leveled principles. These practices are basic when the contemporary condition is sketchy, rough, and genuinely engaged. Firms that can perceive monetary circumstances (e.g., intrusions and customer needs) and are masterminded to develop helpful responses (i.e., spry) are most likely going to achieve more critical execution [18]. The current work looks to tentatively test the effect of learning presentation on fi m execution (**Figures 1** and **2**).

ADVANTAGES	DISADVANTAGES
▪ Learning and more prominent	▪ The loss of control of a customer
▪ More noteworthy time and cost efficiencies	▪ Puts a third party in control of one of the business functions with the most impact on customers
▪ Have room schedule-wise, assets, or skill to adjust their gear and frameworks	▪ Institutional knowledge may be lost
▪ Give higher adaptability to geographic dissemination	▪ Handling the shipping functions in-house
▪ Deal with their assets including workforce size, and transform settled expenses into variable costs	▪ Dedicated transportation contracting
▪ Restricted coordination skill to center around their center business	▪ Transportation procurement.

Figure 1.
Few points described under the advantage and disadvantage of 3PL.

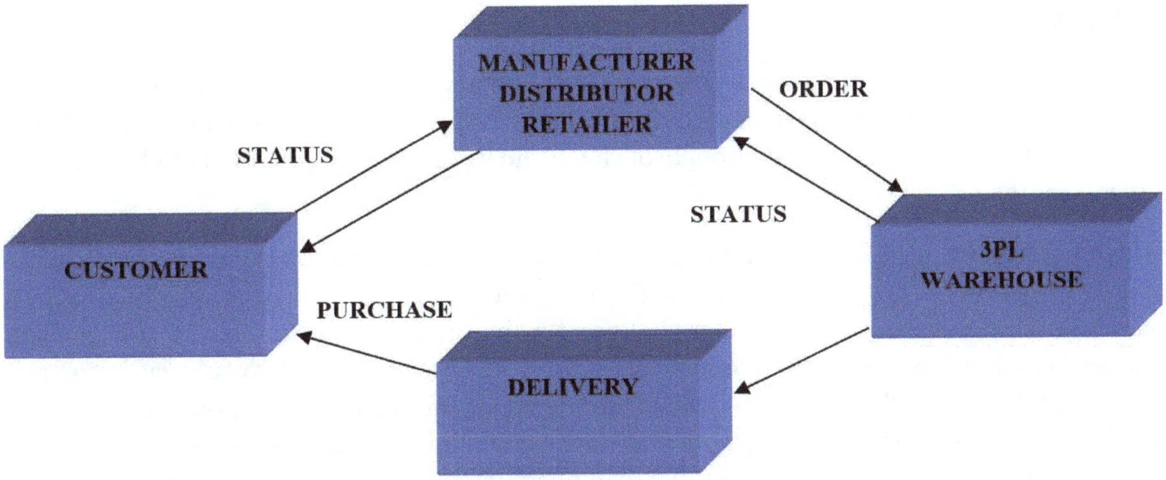

Figure 2.
Cycle from manufacture to distributor to retailer to customer.

The client connector

Th s kind of 3PL supplier comes in line with the client and basically assumes fin sh control of the organization's coordination exercises. The 3PL supplier enhances the coordination drastically yet does not build up another administration. The client base for this kind of 3PL supplier is ordinarily very little. As shown by the social view [33], there are four sources that can make between progressive high grounds: (1) interests in association-specific assets, (2) proficie t plans, (3) complementary resources/capacities, and (4) reasonable organization. In our examination, we joined trust and obligation as the "effective organization" to upgrade 3PL re-appropriating execution. According to [34], the well-developed organization can reduce trade expenses and advance regard creation. In addition, easygoing self-maintaining organization (e.g., trust and obligation) is more suitable than outcast executing government (e.g., legal contracts) or formal self-approving organization (e.g., detainee) in creating social rents. Trust and obligation are seen as basic enabling specialists of productive business associations [89] and can be an advantage prepared for making functional high grounds (e.g., [13–24, 137]). Trust can be portrayed as the conviction that the relationship assistants would not act sagaciously. In the composition, trust between relationship associates has been seen as the most important organization framework [32–57, 59–98, 100–127, 129–133].

One of the huge impediments of creation system and collaborations in organization enhancement in China is the serious inadequacy of qualifi d collaborations and stock system organization authority [124]. Studies demonstrate that an essential test for 3PL fi ms is to select and hold talented specialists [112, 124]. Both worldwide and neighborhood vital pro associations perceive the nonattendance of capacity as one of the key troubles of working in China [30, 62]. The absence of administrators with huge collaboration guidance and getting ready on both conventional and key measurements blocks viability of the 3PL zone in China [134]. In any case, to beat such an issue, worldwide associations rely upon planning, headway, and training to vital close-by capacities, instead of utilizing banishes with high costs [134]. All people in the store have unique association culture, which isolates them from each other [52–57, 59–65, 142]. In any case, seeing each other's lifestyle and various leveled characteristics is a fundamental factor to be engaged in business [120]. Agents of multinationals working in China should think about adjacent culture [97]. Social confli ts frequently result in changed solicitations, unexpected costs, and on occa- sion a negative impact on associations [65].

The client designer

Th s is the most abnormal amount that a 3PL supplier can achieve as for its procedures and exercises. Th s happens when the 3PL supplier coordinates itself with the client and assumes control over their whole coordination work. These suppliers will have couple of clients, however, and will perform broad and nitty-gritty assignments for them.

Redistributing may include a subset of a task's coordination, abandoning a few items or working advances immaculate in light of the fact that the in-house coordination can take the necessary steps preferable or less expensive over an outside provider. Another vital point is the client introduction of the 3PL supplier. The supplier needs to fit to the structures and the necessities of the organization. Th s could easily compare the pure investment funds of 3PL suppliers for plain demonstration. The client introduction in type of versatility to changing client needs, unwavering quality and the adaptability of outsider coordination supplier were made reference to as substantially more essential than unadulterated cost reserve funds. The proposed MN-3PL challenge framework contains a sorted out dynamic arrangement of test orders and troubles. To perceive the fundamental challenges of MN3PL, in the proposed structure, boss needs to consign passionate weights for judgments.

Additionally, since AHP is prepared for overseeing emotional parts of criteria with dynamic judgment [88], the use of AHP as an examination strategy is an appropriate philosophy for recognizing fundamental challenges looked by MN3PLs working in China. AHP does not require incalculable for examination. Clearly, AHP is the typical responses of senior heads and authorities to fi d out the issues under investigation [71, 73] and the eventual outcomes of the AHP-based examination are not affected by a little precedent gauge [104, 113].

Coordination benefit viability is characterized as the degree to which conveying coordination benefit is proficie t [94]. Powerful administration may incorporate on-time conveyance, convenient reaction to demands, exact data stockpiling, and critical thinking [16–57, 59–86], which may eventually expand business execution [67]. In this manner, we suggest that a 3PL's trust in and pledge to key clients and learning introduction all affect its coordination benefit adequacy.

With the collaboration between learning introduction, trust, and responsibility from the dynamic ability point of view, authoritative learning is a nonstop powerful procedure of misusing interior and outside assets to more readily adapt to the changing condition [125]. In this procedure, learning associations will in general develop great associations with their accomplices with the end goal to advance shared alteration, accordingly expanding their trust in and promise to accomplices [106]. From the social view, the mix of assets' cross-associations needs the help of dynamic abilities, particularly organization capacities or cooperation elements through learning and coordination [7–22, 33, 39].

Trust, responsibility, and administrative advancement of a cozy connection between social partners enhance execution [53, 84]. Trusting and submitted connections are especially vital for 3PLs on the grounds that they may lessen wasteful aspects caused from advantage and resource specific ty [138]. Firms might be persuaded to volunteer trust and responsibility as a result of the possibilities of accepting expanded, joint adjustments from collaboration [48]. Social trust and duty additionally flag a company's accentuation on esteem creation through authoritative joint efforts as opposed to protections keeping the astute conduct. It increases the personal circumstance stake and accomplishment between reliance [48]. Authoritative trust and responsibility empower a regenerative cycle decreasing advantage and upgrading collaboration. As the social view proposes, trust and duty are casual self-implementing shields that may enhance hierarchical execution.

The principle predecessor of full of feeling duty is trust, as exhibited by an accomplice association's straightforward and generous conduct [45]. For instance, Morgan and Hunt [89] contended that no responsibility could be culminated except if the accomplices feel the foundation of whole trust. In the profoundly associated inventory network setting, Kwon and Suh [68, 69] recognized trust as main determinant of duty. On the off chance that a gathering feels that a relationship will bolster its interests and react to its needs, it will probably rely upon, and cling to, the relationship [74]. In this manner, we propose the following: a 3PL company's trust in its key clients is emphatically identifi d with its responsibility to its key clients. Connection of learning and trust/responsibility as critical segments may likewise direct the impact of learning introduction on execution. We contend that larger amounts of trust and duty could fortify the connection between learning introduction and coordination benefit viability. On the other hand, bringing down dimensions of trust and duty may baffl the connection between learning introduction and compelling coordination execution.

Without trust, volitional data traded between social accomplices might be incorrect [31]. One extraordinary doubt may pollute data shared between associations [95]. At least an absence of trust may cause wasteful correspondence. Data trade may make chance for the accomplishment of giving data since learning overfl w may happen [56]. Data asymmetry may debilitate one social accomplice's situation by making a lopsidedness of intensity. Without the capacity to confide in their accomplice, sharing data puts the information proprietor at a focused detriment [28]. Expanded trust and duty enhance correspondence exactness prompting enhanced execution. For instance, Inkpen and Currall [56] contended that trust is "a key variable and a type of cash that decides learning openness." The dynamic capacity infers that learning instruments can create dynamic abilities in this way giving an upper hand, and the social view proposes that trust and responsibility as interorganizational linkage are a wellspring of upper hands.

3PL suppliers center their business activities to re-appropriate their coordination exercises for coordination benefits. Re-appropriating gives a specific power that is not accessible inside an association's inward divisions. Th s power can have numerous measurements: economies of scale, process ability, access to capital, access to costly innovation, and so forth. By re-appropriating coordination exercises, fi ms can save money on capital venture and along these lines diminish budgetary dangers [3, 5]. Numerous 3PL organizations give an expert coordination benefit. In the worldwide store network frameworks, ventures attempt to re-appropriate the coordination. 3PL suppliers are one of the decisions. Picking the 3PL suppliers giving the best choice issues is a fascinating and imperative subject of organizations with face when endeavoring to choose a reasonable and long haul 3PL organization. Aguezzol [4] presented a clear picture of 3PL choices focusing on criteria and strategies. In light of the examination of 67 articles distributed between 1994 and 2013, this investigation recognized 11 entry criteria as far as 3PL determination. Cost is the most generally embraced rule, trailed by relationship, administrations, quality, data and hardware framework, adaptability, conveyance, polished methodology, money-related position, area, and notoriety, and as far as strategies for 3PL assessment are concerned, 5 bunches were distinguished to be specific: MCDM systems, measurable methodologies, man-made consciousness, numerical programming, and half and half techniques. Numerous chiefs or specialists select providers dependent on their experience and instinct. These methodologies are clearly emotional. Then again, various criteria basic leadership or different properties basic leadership (MCDM/MADM) is the methodology managing the positioning and determination of at least one provider from a pool of suppliers. The MCDM gives a powerful structure to provider correlation depending on the assessment of numerous contention criteria [114]. One of the strategies for taking care of MCDM issues is explanatory pecking order process (AHP). AHP is an abstract instrument with which to examine, in light of a fresh 9-point scale, the subjective criteria expected to produce elective needs and inclinations. AHP empowers chiefs to develop complex issues in a straightforward various leveled shape and to assess a substantial number of quantitative and subjective factors in a precise way regardless of the nearness of numerous clashing criteria.

Advantages

Cost and time savings

Coordination is the center fitness of outsider coordination suppliers. Suppliers may have better related learning and more prominent skill than the delivering or offering organization and may likewise have more worldwide systems empowering more noteworthy time and cost effici cies.

The gear and the IT frameworks of 3PL suppliers are always refreshed and adjusted to coordinate the prerequisites of their clients and their clients' providers. Delivering or offering organizations frequently do not have room schedule-wise or for assets or skill to adjust their gear and frameworks as quickly [117].

Low capital responsibility

Assuming most or every single agent work is redistributed to a 3PL supplier, there is typically no requirement for the customer to possess its very own stock-room or transport offices bringing down the measure of capital required for the customer's matter of fact. Th s is especially helpful if an organization's stockroom has high varieties in limit use, prompting over purchasing of distribution center limit and lessening productivity.

Core interest

Coordination redistributing enables organizations with restricted coordination skill to center around their center business. Expanding unpredictability in business proposes that organizations profit by not committing assets to regions in which they are not skilled [117].

Adaptability

Outsider coordination suppliers can give higher adaptability to geographic dissemination and may offer a bigger assortment of administrations than customers could accommodate themselves. Th s additionally enables organizations to typically deal with their assets including workforce size and to transform settled expenses into variable costs [118].

Disadvantages

Thi d party logistics sounds like a slam dunk, does not it? Well, before you sign on the dotted line with your friendly, local 3PL rep, let us look at both sides of the equation. One major disadvantage of using 3PL is that it leads to some loss of control over your shipping functions. Th s business strategy puts a third party in control of one of the business functions with the most impact on customers and the greatest effect on customer satisfaction. In addition, turning these functions over to a 3PL fi m is a major commitment, since your in-house logistical team may lose much of the relevant market knowledge that you will need should the relationship with your 3PL provider ever become untenable. Bringing the shipping function back in-house could prove difficult when this institutional knowledge is lost. The cost factor is not crystal clear either. A 3PL fi m may be cheaper up front, but over time, it will likely be more expensive than handling the shipping functions in-house, provided those in-house functions are operating effici tly.

Loss of control

One drawback is the loss of control a customer hosts by utilizing third gathering coordination. With outbound coordination, the 3PL supplier for the most part accepts correspondence and cooperates with a company's client or provider. To alleviate this, some 3PLs endeavor to mark themselves as their customers, for example, applying customers' logos on their advantages and dressing their representatives like their customers' employees [115].

Conclusion

Thi d-party logistics was propelled to bear the weight of obligations from various organizations. At fi st, it had distinctive jobs to perform, yet its prosperity and viability supported worldwide organizations as well as it lifted the measures and productivity of various organizations in the aggressive market. Lead coordination suppliers however do not have quite a bit of their benefits yet their skill in the fi ld makes them champion. They have low overhead costs, great correspondence, and compelling investigation with their solid system framework. Their proficie cy and affectivity in their fi lds are strikingly extraordinary because of their group's potential. At the point when outsider coordination is contrasted and second gathering coordination, it demonstrates the distinction unmistakably. Second gathering coordination does not have an incorporated framework and edge work for their dealings and administration giving assignments. The outstanding task at hand of outsider coordination is very high as they oblige future objectives and undertakings. In addition, there have been diverse ways to deal with and think about the viability of outsider coordination, yet none of the proposed model or procedure suits the best element for questioning outsider coordination.

Author details

Yangyan Shi*, Rafay Waseem and Hafiz uhammad Shahid
Department of Economics and Management, College of Agricultural Economics and Management, Nanjing Agricultural University, Nanjing, China

*Address all correspondence to: peter.shi@mq.edu.au

References

[1] 11 Ways to Gain Global Transport Cost Control. ControlPay—Global Processing of Transport Data. Retrieved 21 January 2018

[2] Supply Chain Visibility Is Ranked Top Priority. 22 July 2013. Retrieved 21 January 2018

[3] Aghazadeh SM. How to choose an effective third party logistics provider. Management Research News. 2003;**26**(7):50-58

[4] Aguezzoul A. Thi d-party logistics selection problem: A literature review on criteria and methods. Omega. 2014;**49**:69-78

[5] Aktas E, Ulengin F. Outsourcing logistics activities in Turkey. Journal of Enterprise Information Management. 2005;**18**(3):316-329

[6] Alkhatib SF, Darlington R, Yang Z, Nguyen TT. A novel technique for evaluating and selecting logistics service providers based on the logistics resource view. Expert Systems with Applications. 2015;**42**(20):6976-6989

[7] Allred CR, Fawcett SE, Wallin C, Magnan GM. A dynamic collaboration capability as a source of competitive advantage. Decision Sciences. 2011;**42**(1):129-161

[8] Andersson D. Thi d Party Logistics: Outsourcing Logistics in Partnerships. Department of Management and Economics: Linköping University; 1997

[9] Heck A. Strategische Allianzen. Springer Verlag; pp. 47+48

[10] Attaran M. RFID: An enabler of supply chain operations. Supply Chain Management. 2007;**12**(4):249-257

[11] Azadegan A, Dooley KJ. Supplier innovativeness, organizational learning styles and manufacturer performance: An empirical assessment. Journal of Operations Management. 2010;**28**(6):488-505

[12] Bagchi PK, Chun Ha B, Skjoett- Larson T, Soerensen LB. Supply chain integration: A European survey. International Journal of Logistics Management. 2005;**16**(2):275-294

[13] Barney JB, Clark DN. Resource- Based Theory: Creating and Sustaining Competitive Advantage. Oxford. New York: Oxford University Press; 2007

[14] Bartolacci MR, Leblanc LJ, Kayikci Y, Grossman TA. Optimization modeling for logistics: Options and implementations. Journal of Business Logistics. 2012;**33**(2):118-127

[15] Berglund M, van Laarhoven P, Sharman G, Wandel S. Thi d-party logistics: Is there a future? The International Journal of Logistics Management. 1999;**10**(1):59-70

[16] Bienstock CC, Mentzer JT, Bird MM. Measuring physical distribution service quality. Journal of the Academy of Marketing Science. 1997;**25**(1):31-44

[17] Bottani E, Rizzi A. A fuzzy TOPSIS methodology to support outsourcing of logistics services. Supply Chain Management: An International Journal. 2006;**11**(4):294-308

[18] Braunscheidel MJ, Suresh NC. The organizational antecedents of a fi m's supply chain agility for risk mitigation and response. Journal of Operations Management. 2009;**27**(2):119-140

[19] Bustinza OF, Arias-Aranda D, Gutierrez-Gutierrez L. Outsourcing, competitive capabilities and performance: An empirical study in service fi ms. International Journal of Production Economics. 2010;**126**(2):276-288

[20] Cao M, Zhang Q. Supply chain collaboration: Impact on collaborative advantage and fi m performance. Journal of Operations Management. 2011;**29**(3):163-180

[21] Carter CR. Purchasing social responsibility and fi m performance: The key mediating roles of organizational learning and supplier performance. International Journal of Physical Distribution and Logistics Management. 2005;**35**(3):177-194

[22] Chang KH, Chen YR, Huang HF. Information technology and partnership dynamic capabilities in international subcontracting relationships. International Business Review. 2015;**24**(2):276-286

[23] Chen KY, Wu WT. Applyıng analytic network process in logistics service provider selection—A case study of the industry investing in Southeast Asia. International Journal of Electronic Business Management. 2011;**9**(1):24-36

[24] Chen YH, Lin TP, Yen DC. How to facilitate inter-organizational knowledge sharing: The impact of trust. Information and Management. 2014;**51**(5):568-578

[25] Cheung MS, Myers MB, Mentzer JT. The value of relational learning in global buyer-supplier exchanges: A dyadic perspective and test of the pie- sharing premise. Strategic Management Journal. 2011;**32**(10):1061-1082

[26] Chin FC, Bae J-H, Kim GO. A survey on the logistics service providers in Shanghai. International Journal of Physical Distribution & Logistics Management. 2007;**29**(9):588-605

[27] Chu Z. Logistics and economic growth: A panel data approach. The Annals of Regional Science. 2012;**49**(1):87-102

[28] Cohen WM, Levinthal DA. Absorptive capacity: A new perspective on learning and innovation. Administrative Science Quarterly. 1990:128-152

[29] Crawley D. Is your 3PL fi ancially healthy?. 2013. http://www. inboundlogistics.com/cms/article/is-your-3pl-fi ancially-healthy/ [Accessed: 10 October 2016]

[30] Cui L, Su S-II, Hertz S. Logistics innovation in China. Transportation Journal. 2012;**51**(1):98-117

[31] Currall SC, Judge TA. Measuring trust between organizational boundary role persons. Organizational Behavior and Human Decision Processes. 1995;**64**(2):151-170

[32] Dyer JH, Chu W. The determinants of trust in supplier-automaker relationships in the U.S., Japan, and Korea. Journal of International Business Studies. 2000;**31**(2):259-285

[33] Dyer JH, Singh H. The relational view: Cooperative strategy and sources of inter organizational competitive advantage. Academy of Management Review. 1998;**23**(4):660-679

[34] Ebrahim-Khanjari N, Hopp W, Iravani SM. Trust and information sharing in supply chains. Production and Operations Management. 2012;**21**(3):444-464

[35] Ecer F. Thi d-party logistics (3PLs) provider selection via fuzzy AHP and EDAS integrated model. Technological and Economic Development of Economy. 2018;**24**(2):615-634

[36] Eisenhardt KM, Martin JA. Dynamic capabilities: What are they? Strategic Management Journal. 2000:211105-211121

[37] Falsini D, Fondi F, Schiraldi MM. A logistics provider evaluation and selection methodology based on AHP, DEA and linear programming integration. International Journal of Production Research. 2012;**50**(17):4822-4829

[38] Fang EA, Li X, Lu J. Effects of organizational learning on process technology and operations performance in mass customizers. International Journal of Production Economics. 2016:17468-17475

[39] Fawcett SE, Fawcett AM, Watson BJ, Magnan GM. Peeking inside the black box: Toward an understanding of supply chain collaboration dynamics. Journal of Supply Chain Management. 2012;**48**(1):44-72

[40] Fifth Party Logistic Model (5PL). Logistics Glossary. Retrieved 21 September 2018

[41] Frazelle E. Supply Chain Strategy: The Logistics of Supply Chain Management. McGraw Hill; 2002

[42] Fu B, Bentz BA, McCalla MT. Logistics in China: Thi k-ing ahead. Logistics Management. 2011;**50**(10):36-40

[43] Fu S, Han Z, Huo B. Relational enablers of information sharing: Evidence from Chinese food supply chains. Industrial Management & Data Systems. 2017;**117**(5):838-852

[44] García-Morales VJ, Jiménez- Barrionuevo MM, Gutiérrez-Gutiérrez L. Transformational leadership influence on organizational performance through organizational learning and innovation. Journal of Business Research. 2012;**65**(7):1040-1050

[45] Geyskens I, Steenkamp J, Scheer L, Kumar N. The effects of trust and interdependence on relationship commitment: A trans-Atlantic study. International Journal of Research in Marketing. 1996;**13**(4):303-317

[46] Girod SJ, Whittington R. Reconfigur tion, restructuring and fi m performance: Dynamic capabilities and environmental dynamism. Strategic Management Journal. 2017;**38**(5):1121-1133

[47] Govindan K, Chaudhuri A. Interrelationships of risks faced by third party logistics service providers: A DEMATEL based approach. Transportation Research Part E: Logistics and Transportation Review. 2016;**90**:177-195

[48] Gundlach GT, Achrol RS, Mentzer JT. The structure of commitment in exchange. Journal of Marketing. 1995;**59**(1):78-92

[49] Hensher DA, Zhang Z, Rose JM. Logistics Challenges for China: Drivers of the Logistics Industry Growth, and Bottlenecks Constraining Development. University of Sydney, Australia: Institute of Transport and Logistics Studies; 2015

[50] Hertz S, Alfredsson M. Strategic development of third party logistics providers. Industrial Marketing Management Elsevier Science. February 2003;**32**(2):139-149. DOI: 10.1016/ S0019-8501(02)00228-6

[51] Hobbs JE. A transaction cost approach to supply chain management. Supply Chain Management: An International Journal. 1996;**1**(2):15-27

[52] Hong J, Chin AT, Liu B. Logistics outsourcing by manufacturers in China: A survey of the industry. Transportation Journal. 2004:17-25

[53] Hult GTM, Ketchen DJ, Nichols EL. Organizational learning as a strategic resource in supply management. Journal of Operations Management. 2003;**21**(5):541-556

[54] Huo B, Ye Y, Zhao X. The impacts of trust and contracts on opportunism in the 3PL industry: The moderating role of demand uncertainty. International Journal of Production Economics. 2015:170160-170170

[55] Hwang BN, Shen YC. Decision making for third party logistics supplier selection in semiconductor manufacturing industry. A nonadditive fuzzy integral approach. Mathematical Problems in Engineering. 2015;918602 12 pages

[56] Inkpen AC, Currall SC. The coevolution of trust, control, and learning in joint ventures. Organization Science. 2004;**15**(5):586-599

[57] Jiao Z. Operations mode and development trend of online shopping logistics in China. Contemporary Logistics in China. Springer; 2014

[58] JOC. Logistics benefiting from China's rapid e-commerce growth. Journal of Commerce. 2014:213-

233. Available from: http://www. joc.com/international-logistics/ logisticsbenefiting-china%E2%80%99s- rapid-e-commerce-growth_20141216. html [Accessed: 17 April 2017]

[59] Joo S-J, Keebler JS, Hanks S. Measuring the longitudinal performance of 3PL branch operations. Benchmarking: An International Journal. 2013;**20**(2):251-262

[60] Jordan D. Fourth party logistics (4PLP); what this means for your supply chain. Journal of Marketing. 2010;**58**(3):20-38. Exchange. Retrieved: 21 September 2018

[61] Jung H. Evaluation of third party logistics providers considering social sustainability. Sustainability. 2017;**9**(5):777. 18 pages

[62] Kam B, Tsahuridu E, Ding M. Does human resource management contribute to the development of logistics and supply chain capabilities? An empirical study of logistics service providers in China. In: 11th International Conference of the Society for Global Business & Economic Development; Montclair State University. 2009

[63] Kannan G, Pokharel S, Sasi Kumar P. A hybrid approach using ISM and fuzzy TOPSIS for the selection of reverse logistics provider. Resources, Conservation and Recycling. 2009;**54**:28-36

[64] Kash BA, Spaulding A, Gamm LD, Johnson CE. Healthcare strategic management and the resource based view. Journal of Strategy and Management. 2014;**7**(3):251-264

[65] Kerr J. 10 key challenges for the Chinese logistics industry. Logistics Management. 2005;**44**(2):64-68

[66] KeshavarzGhorabaee M, Amiri M, Zavadskas EK, Antucheviciene J. Assessment of third-party logistics providers using a CRITIC–WASPAS approach with interval type-2 fuzzy sets. Transport. 2017;**32**(1):66-78

[67] Knemeyer AM, Murphy PR. Evaluating the performance of third-party logistics arrangements: A relationship marketing perspective. Journal of Supply Chain Management: A Global Review of Purchasing & Supply. 2004;**40**(1):35-51

[68] Kwon IG, Suh T. Factors affecting the level of trust and commitment in supply chain relationships. Journal of Supply Chain Management: A Global Review of Purchasing and Supply. 2004;**40**(2):4-14

[69] Kwon IG, Suh T. Trust, commitment and relationships in supply chain management: A path analysis. Supply Chain Management: An International Journal. 2005;**10**(1):26-33

[70] Lai F, Tian Y, Huo B. Relational governance and opportunism in logistics outsourcing relationships: Empirical evidence from China. International Journal of Production Research. 2012;**50**(9):2501-2514

[71] Lam K, Zhao X. An application of quality function deployment to improve the quality of teaching. International Journal of Quality & Reliability Management. 1998;**15**(4):389-413

[72] Leahy S, Murphy P, Poist R. Determinants of successful logistics relationships: A third party provider perspective. Transportation Journal. 1995;**35**:5-13

[73] Lee S, Ross SD. Sport sponsorship decision making in a global market. Sport, Business and Management: An International Journal. 2012;**2**(2):156-168

[74] Li D, Browne GJ, Chau PYK. An empirical investigation of web site use using a commitment- based model. Decision Sciences. 2006;**37**(3):427-444

[75] Lieb RC. The use of third-party logistics services by large American manufacturers. Journal of Business Logistics. 1992;**13**(2):29-42

[76] Lin Y, Wu LY. Exploring the role of dynamic capabilities in fi m performance under the resource-based view framework. Journal of Business Research. 2014;**67**(3):407-413

[77] Liu HT, Wang WK. An integrated fuzzy approach for provider evaluation and selection in third-party logistics. Expert Systems with Applications. 2009;**36**:4387-4398

[78] Liu JJ, So SC, Choy KL, Lau H. Performance improvement of third-party logistics providers? An integrated approach with a logistics information system. International Journal of Technology Management. 2008;**42**(3):226-249

[79] Liu X. China-based logistics research: A review of the literature and implications. International Journal of Physical Distribution & Logistics Management. 2014;**44**(5):392-411

[80] Lohman C, Fortuin L, Wouters M. Designing a performance measurement system: A case study. European Journal of Operational Research. 2004;**156**:267-286. DOI: 10.1016/S0377-2217(02)00918-9

[81] Mahpula A, Yang D, Kurban A, Witlox F. An overview of 20 years of Chinese logistics research using a content-based analysis. Journal of Transport Geography. 2013;**31**:30-34

[82] Marasco A. Thi d-party logistics: A literature review. International Journal of Production Economics. 2008;**113**(1):127-147

[83] Murray M. Selecting a Thi d Party Logistics (3PL) Provider. Thebalance.com

[84] Martin JH, Grbac B. Using supply chain management to leverage a fi m's market orientation. Industrial Marketing Management. 2003;**32**(1):25-38

[85] Mentzer JT, Flint DJ, Hult GTM. Logistics service quality as a segment customized process. Journal of Marketing. 2001;**65**(4):82-104

[86] Mentzer JT, Gomes R, Krapfel RE. Physical distribution service: A fundamental marketing concept? Journal of the Academy of Marketing Science. 1989;**17**(1):53-62

[87] Millar M. China Logistics Sector Developments. 2014. Available from: http://www.eft.c m/chinalogistics- sector-developments [Accessed: 18 August 2016]

[88] Perçin S. Evaluation of third-party logistics (3PL) providers by using a two- phase AHP and TOPSIS methodology. Benchmarking: An International Journal. 2009;**16**(5):588-604

[89] Morgan RM, Hunt SD. The commitment-trust theory of relationship marketing. Journal of Marketing. 1994;**58**(3):20-38

[90] Schröter N, Schröter I. Supply Chain Management und Logistik. Stuttgart: Verlag W. Kohlhammer; 2010, Seite 15

[91] Özer Ö, Zheng Y, Ren Y. Trust, trustworthiness, and information sharing in supply chains bridging China and the United States. Management Science. 2014;**60**(10):2435-2460

[92] Pamucar D, Mihajlovic M, Obradovic R, Atanaskovic P. Novel approach to group multi-criteria decision making based on interval rough numbers: Hybrid DEMATEL-ANP-MAIRCA model. Expert Systems with Applications. 2017;**88**(1):58-80

[93] Pamucar D, Stevic E, Zavadskas EK. Integration of interval rough AHP and interval rough MABAC for evaluating university web pages. Applied Soft Computing. 2018;**67C**:141-163

[94] Panayides PM. The impact of organizational learning on relationship orientation, logistics service effectiveness and performance. Industrial Marketing Management. 2007;**36**(1):68-80

[95] Powell WW, Koput KW, Smith-Doerr L. Interorganizational collaboration and the locus of innovation: Networks of learning in biotechnology. Administrative Science Quarterly. 1996;**41**:116-145

[96] Qureshi MN, Kumar D, Kumar P. An integrated model to identify and classify the key criteria and their role in the assessment of 3PL services providers. Asia Pacific Journal of Marketing and Logistics. 2008;**20**(2):227-249

[97] Rahman S, Wu Y-CJ. Logistics outsourcing in China: The manufacturer-cumsupplier perspective. Supply Chain Management: An International Journal. 2011;**16**(6):462-473

[98] Rahman Z. Use of internet in supply chain management: A study of Indian companies. Industrial Management & Data Systems. 2004;**104**(1):31-41

[99] Cheong ML. Logistics outsourcing and 3PL challenges. 2004. http://hdl. handle.net/1721.1/3908 [Accessed: 14 April 2017]

[100] Rai A, Pavlou PA, Im G, Du S. Interfi m IT capability profiles and communications for cocreating relational value: Evidence from the logistics industry. MIS Quarterly. 2012;**36**(1):233-262

[101] Ramaa A, Rangaswamy TM, Subramanya KN. A review of literature on performance measurement of supply chain network. In: 2009 2nd International Conference on Emerging Trends in Engineering and Technology (ICETET). 2009. pp. 802-807. DOI: 10.1109/ICETET.2009.18

[102] Rushton A, Croucher P, Baker P. The Handbook of Logistics & Distribution Management. 4th ed. UK: Kogan Page Limited/The Chartered Institute of Logistics and Transport; 2010

[103] Sahu NK, Datta S, Mahapatra SS. Fuzzy based appraisement module for 3PL evaluation and selection. Benchmarking: An International Journal. 2015;**22**(3):354-392

[104] Sambasivan M, Fei NY. Evaluation of critical success factors of implementation of ISO 14001 using analytic hierarchy process (AHP): A case study from Malaysia. Journal of Cleaner Production. 2008;**16**(13):1424-1433

[105] Santos-Vijande ML, López-Sánchez JÁ, Trespalacios JA. How organizational learning affects a fi m's flex bility, competitive strategy, and performance. Journal of Business Research. 2012;**65**(8):1079-1089

[106] Santos-Vijande ML, Sanzo-Pérez MJ, Álvarez-González LI, Vázquez- Casielles R. Organizational learning and market orientation: Interface and effects on performance. Industrial Marketing Management. 2005;**34**(3):187-202

[107] Selviaridis K, Spring M. Thi d party logistics: A literature review and research agenda. The International Journal of Logistics Management. 2007;**18**(1):125-150

[108] Senthil S, Srirangacharyulu B, Ramesh A. A robust hybrid multi- criteria decision making methodology for contractor evaluation and selection in third-party reverse logistics. Expert Systems with Applications. 2014;**41**:50-58

[109] Sha M, Guan C. Characteristics and trends of the third party logistics market in mainland China. Logistics: The Emerging Frontiers of Transportation and Development in China. ASCE; 2009. pp. 1716-1723

[110] Shanghai International Journal of Physical Distribution & Logistics Management;**29**(9):588-605

[111] Sharma SK, Kumar V. Optimal selection of third-party logistics service providers using quality function deployment and Taguchi loss function. Benchmarking: An International Journal. 2015;**22**(7):1281-1300

[112] Shi Y, Zhang A, Arthanari T, Liu Y, Cheng TCE. Thi d-party purchase: An empirical study of third-party logistics providers in China. International Journal of Production Economics. 2016;**171**(3):189-200

[113] Shrestha RK, Alavalapati JR, Kalmbacher RS. Exploring the potential for silvopasture adoption in south- central Florida: An application of SWOT–AHP method. Agricultural Systems. 2004;**81**(3):185-199

[114] Shyur HJ, Shih HS. A hybrid MCDM model for strategic vendor selection. Mathematical and Computer Modelling. 2006;**44**(7):749-761

[115] Simchi-Levi, Kaminsky. Designing and Managing the Supply Chain: Concepts, Strategies and Case Studies. 3rd ed. McGraw-Hill International Edition. p. 252

[116] Simchi-Levi, Kaminsky. Designing and Managing the Supply Chain: Concepts, Strategies and Case Studies. 3rd ed. McGraw-Hill International Edition. p. 253

[117] Simchi-Levi, Kaminsky. Designing and Managing the Supply Chain: Concepts, Strategies and Case Studies. 3rd ed. McGraw-Hill International Edition. p. 251

[118] Simchi-Levi, Kaminsky. Designing and Managing the Supply Chain: Concepts, Strategies and Case Studies. 3rd ed. McGraw-Hill International Edition. p. 250

[119] Skjoett-Larsen T. Thi d party logistics—From an interorganizational point of view. International Journal of Physical Distribution & Logistics Management. 2000;**30**(2):112-127

[120] Sohal AS, Rahman S. Use of third party logistics services: An Asia-Pacific perspective. In: Handbook of Global Logistics. Springer; 2013. pp. 45-67

[121] Song H, Wang L. The status and development of logistics cost

management: Evidence from mainland China. Benchmarking: An International Journal. 2009;**16**(5):657-670

[122] Song S, Fan M. Thi d-party logistics information system security measures. In: International Conference on Transportation Engineering 2009. ASCE; 2009

[123] Su HC, Chen YS. Unpacking the relationships between learning mechanisms, culture types, and plant performance. International Journal of Production Economics. 2013;**146**(2):728-737

[124] Tan AWK, Yifei Z, Zhang D, Hilmola O-P. State of third party logistics providers in China. Industrial Management & Data Systems. 2014;**114**(9):1322-1343

[125] Teece DJ, Pisano G, Shuen A. Dynamic capabilities and strategic management. Strategic Management Journal. 1997:509-533

[126] Tian Y, Ellinger AE, Chen H. Thi d-party logistics provider customer orientation and customer fi m logistics improvement in China. International Journal of Physical Distribution & Logistics Management. 2010;**40**(5):356-376

[127] Vitasek K. Council of S. C. M. P. Supply Chain Management Terms and Glossary; 2013

[128] Christopher M. Logistics and supply chain management—Strategies for reducing cost and improving service. In: Financial Times. 3rd ed. Prentice Hall; 2005

[129] Vogel R, Güttel WH. The dynamic capability view in strategic management: A bibliometric review. International Journal of Management Reviews. 2013;**15**(4):426-446

[130] Wallenburg CM, Cahill DL, Michael Knemeyer A, Goldsby TJ. Commitment and trust as drivers of loyalty in logistics outsourcing relationships: Cultural differences between the United States and Germany. Journal of Business Logistics. 2011;**32**(1):83-98

[131] Waters D. Logistics: An introduction to supply chain management. Supply Chain Management: An International Journal. 2003:364

[132] Weber C, Bauke B, Raibulet V. An empirical test of the relational view in the context of corporate venture capital. Strategic Entrepreneurship Journal. 2016;**10**(3):274-299

[133] Weber C, Weidner K, Kroeger A, Wallace J. Social value creation in inter organizational collaborations in the not-for-profit sector—Give and take from a dyadic perspective. Journal of Management Studies. 2017;**54**(6):929-956

[134] Wilkinson B, Eberhardt M, McLaren J, Millington A. Human resource barriers to partnership sourcing in China. The International Journal of Human Resource Management. 2005;**16**(10):1886-1900

[135] Williamson OE. The economic institutions of capitalism. 1985

[136] Williamson OE. Economics and organization: A primer. California Management Review. 1996;**38**:131-146

[137] Wu F, Cavusgil ST. Organizational learning, commitment, and joint value creation in interfi m relationships. Journal of Business Research. 2006;**59**(1):81-89

[138] Wu F, Sinkovics R, Cavusgil ST, Roath AS. Overcoming export manufacturers' dilemma in international expansion. Journal of International Business Studies. 2007;**38**(2):283-302

[139] Wu F, Yeniyurt S, Kim D, Cavusgil ST. The impact of information technology on supply chain capabilities and fi m performance: A resource- based view. Industrial Marketing Management. 2006;**35**(4):493-504

[140] Yang H, Zheng Y, Zaheer A. Asymmetric learning capabilities and stock market returns. Academy of Management Journal. 2015;**58**(2):356-374

[141] Yayla AY, Oztekin A, Gumuş AT, Gunasekaran A. A hybrid data analytic methodology for 3PL transportation provider evaluation using fuzzy multi- criteria decision making. International Journal of Production Research. 2015;**53**(20):6097-6113

[142] Yeung JHY, Selen W, Sum C-C, Huo B. Linking fi ancial performance to strategic orientation and operational priorities: An empirical study of third-party logistics providers. International Journal of Physical Distribution & Logistics Management. 2006;**36**(3):210-230

[143] Zacharia ZG, Sanders NR, Nix NW. The emerging role of the third- party logistics provider (3PL) as an orchestrator. Journal of Business Logistics. 2011;**32**(1):40-54

[144] Zhang G, Shang J, Li W. An information granulation entropy- based model for third-party logistics providers' evaluation. International Journal of Production Research. 2012;**50**(1):177-190

Permissions

All chapters in this book were first published in TSAA, by InTech Open; hereby published with permission under the Creative Commons Attribution License or equivalent. Every chapter published in this book has been scrutinized by our experts. Their significance has been extensively debated. The topics covered herein carry significant findings which will fuel the growth of the discipline. They may even be implemented as practical applications or may be referred to as a beginning point for another development.

The contributors of this book come from diverse backgrounds, making this book a truly international effort. This book will bring forth new frontiers with its revolutionizing research information and detailed analysis of the nascent developments around the world.

We would like to thank all the contributing authors for lending their expertise to make the book truly unique. They have played a crucial role in the development of this book. Without their invaluable contributions this book wouldn't have been possible. They have made vital efforts to compile up to date information on the varied aspects of this subject to make this book a valuable addition to the collection of many professionals and students.

This book was conceptualized with the vision of imparting up-to-date information and advanced data in this field. To ensure the same, a matchless editorial board was set up. Every individual on the board went through rigorous rounds of assessment to prove their worth. After which they invested a large part of their time researching and compiling the most relevant data for our readers.

The editorial board has been involved in producing this book since its inception. They have spent rigorous hours researching and exploring the diverse topics which have resulted in the successful publishing of this book. They have passed on their knowledge of decades through this book. To expedite this challenging task, the publisher supported the team at every step. A small team of assistant editors was also appointed to further simplify the editing procedure and attain best results for the readers.

Apart from the editorial board, the designing team has also invested a significant amount of their time in understanding the subject and creating the most relevant covers. They scrutinized every image to scout for the most suitable representation of the subject and create an appropriate cover for the book.

The publishing team has been an ardent support to the editorial, designing and production team. Their endless efforts to recruit the best for this project, has resulted in the accomplishment of this book. They are a veteran in the field of academics and their pool of knowledge is as vast as their experience in printing. Their expertise and guidance has proved useful at every step. Their uncompromising quality standards have made this book an exceptional effort. Their encouragement from time to time has been an inspiration for everyone.

The publisher and the editorial board hope that this book will prove to be a valuable piece of knowledge for researchers, students, practitioners and scholars across the globe.

List of Contributors

Attila Bai and Zoltán Gabnai
Faculty of Economics and Business, University of Debrecen, Debrecen, Hungary

Jiawen Liu, Yu Wang and Nailing Ge
The Key Laboratory of Road and Traffic Engineering, Ministry of Education, College of Transportation Engineering, Tongji University, Shanghai, China

Hui Li
The Key Laboratory of Road and Traffic Engineering, Ministry of Education, College of Transportation Engineering, Tongji University, Shanghai, China
Department of Civil and Environmental Engineering, University of California Pavement Research Center, University of California, Davis, CA, USA

Corinna Cermak, Erich Markl and Maximilian Lackner
University of Applied Sciences FH Technikum Wien, Vienna, Austria

Francesca Russo, Salvatore Antonio Biancardo and Rosa Veropalumbo
University of Naples Federico II (NA), Italy

Ossama E. Ramadan and Virginia P. Sisiopiku
University of Alabama at Birmingham, Birmingham, AL, USA

Claudia M. Knezek and Susan Polirstok
College of Education, Kean University, NJ, USA

Roxie James
College of Natural and Applied Health Sciences, Kean University, NJ, USA

Anthony Pittman
College of Education, Acting Dean of Education, Kean University, NJ, USA

Gary Poedubicky
College of Education, Acting Dean of Education, Kean University, NJ, USA
Division of Highway Traffic Safety, State of New Jersey, Trenton, NJ, USA

George Tumanishvili, Tengiz Nadiradze and Giorgi Tumanishvili
Institute of Machine Mechanics, Tbilisi, Georgia

Stefano de Luca, Roberta Di Pace and Facundo Storani
University of Salerno, Fisciano, Salerno, Italy

Jozef Gašparík
Faculty of Operation and Economics of Transport and Communications, University of Žilina, Slovakia

Václav Cempírek
College of Logistics, Přerov, Czech Republic

Yangyan Shi, Rafay Waseem and Hafiz Muhammad Shahid
Department of Economics and Management, College of Agricultural Economics and Management, Nanjing Agricultural University, Nanjing, China

Index

www.ingramcontent.com/pod-product-compliance
Lightning Source LLC
LaVergne TN
LVHW080648210926
848485LV00003B/347